Paul Buitenhuis

**The New Competitiveness
in Design and Construction**

THE NEW COMPETITIVENESS

in Design and Construction

by Joe M. Powell

Executive Director
The Rice University Building Institute

John Wiley & Sons, Ltd

Published in Great Britain in 2008 by John Wiley & Sons Ltd

Copyright © 2008 John Wiley & Sons Ltd, The Atrium, Southern Gate, Chichester,
West Sussex PO19 8SQ, England
Telephone +44 (0)1243 779777

Email (for orders and customer service enquiries): cs-books@wiley.co.uk
Visit our Home Page on www.wiley.com

All Rights Reserved. No part of this publication may be reproduced, stored in a retrieval system or transmitted in any form or by any means, electronic, mechanical, photocopying, recording, scanning or otherwise, except under the terms of the Copyright, Designs and Patents Act 1988 or under the terms of a licence issued by the Copyright Licensing Agency Ltd, 90 Tottenham Court Road, London W1T 4LP, UK, without the permission in writing of the Publisher. Requests to the Publisher should be addressed to the Permissions Department, John Wiley & Sons Ltd, The Atrium, Southern Gate, Chichester, West Sussex PO19 8SQ, England, or emailed to permreq@wiley.co.uk, or faxed to +44 (0)1243 770620.

Designations used by companies to distinguish their products are often claimed as trademarks. All brand names and product names used in this book are trade names, service marks, trademarks or registered trademarks of their respective owners. The Publisher is not associated with any product or vendor mentioned in this book.

This publication is designed to provide accurate and authoritative information in regard to the subject matter covered. It is sold on the understanding that the Publisher is not engaged in rendering professional services. If professional advice or other expert assistance is required, the services of a competent professional should be sought.

Other Wiley Editorial Offices

John Wiley & Sons Inc., 111 River Street, Hoboken, NJ 07030, USA

Jossey-Bass, 989 Market Street, San Francisco, CA 94103-1741, USA

Wiley-VCH Verlag GmbH, Boschstr. 12, D-69469 Weinheim, Germany

John Wiley & Sons Australia Ltd, 42 McDougall Street, Milton, Queensland 4064, Australia

John Wiley & Sons (Asia) Pte Ltd, 2 Clementi Loop #02-01, Jin Xing Distripark, Singapore 129809

John Wiley & Sons Canada Ltd, 5353 Dundas Street West, Suite 400, Etobicoke, Ontario M9B 6H8, Canada

Wiley also publishes its books in a variety of electronic formats. Some content that appears in print may not be available in electronic books.

Executive Commissioning Editor: Helen Castle
Project Editor: Miriam Swift
Publishing Assistant: Calver Lezama

ISBN 978-0-470-06560-0

Page design and layouts by Sparks, Oxford – www.sparkspublishing.com
Cover image © istockphoto.com/subman
Cover design by Paul Mitchell Design Ltd
pp. 1, 8, 9, 13, 17, 25, 32, 44, 51, 53, 58, 61, 71, 73, 76, 79, 81, 85, 87, 93, 101, 104, 105, 109, 111, 114, 117, 121, 124, 127, 131, 133, 137, 138, 145, 152, 172,181 & 189: © ImageZoo.com/Veer
pp. 5, 14, 20, 22, 29, 38, 47, 57, 65, 67, 98, 142, 149, 158, 162, 165 & 179: © Photodisc/Getty/Veer
pp. 37, 89, 96, 154, 163, 169, 175 & 187: © Stockbyte/Getty/Veer

Printed and bound by Conti Tipocolor, Italy

CONTENTS

Foreword: A Reflection on Science and Industry   vii
Dedication   x
Acknowledgments   x
Introduction: Shaping the Built Environment   xi

Executive Overview   2
Chapter 1 Vision   10
Chapter 2 Values   30
Chapter 3 Competitive Focus   48
Chapter 4 Category Ownership   62
Chapter 5 Persistent Branding   90
Chapter 6 Marketing Breakthroughs   122
Chapter 7 High-Impact People   146
Chapter 8 New Strategic Relationships   166

Quoted Industry Leaders   191
Index   195

Foreword

A REFLECTION ON SCIENCE AND INDUSTRY

Success in any science or industry relies on a complex mesh of people, technology, and economic conditions. As this book suggests, pursuing the right strategy is one aspect of this complexity. However, it is understood that an essential precondition for strategic success is effective collaboration among a project's various players. In fact, the Rice University Building Institute (RBI) was founded on the belief that collaboration in the huge building industry is so important that it is worthy of serious study. Since this subject is generally not a central focus in studies of this industry, we must turn to interactions in science, particularly to the research of historian Peter Galison, whose history of the working relations among physicists in the 20th century is not only fascinating but also extremely useful. In the end, Galison suggests that the success of the new physics and its players—the theorists, experimenters, and instrument makers—lay in their ability to maintain their individual integrity while forming hybrid coalitions, or trading zones, to achieve the desired scientific breakthroughs.

> I take it to be the sign of vibrant life, not fragility, that the material culture of the laboratory is in flux through changing modes of collaboration, techniques, simulations, and disciplinary alliances.
>
> PETER GALISON
> *IMAGE AND LOGIC: A MATERIAL CULTURE OF MICROPHYSICS*
> UNIVERSITY OF CHICAGO PRESS (1997)

It may be too early to argue that the successful construction of a complex building relies likewise on the general robustness of the industry combined with innovative players able and willing to form hybrid trading zones, but by using Galison's suggestions as a frame of reference, we can begin to see more clearly what takes place in those trading zones. One thing is clear: Zones do exist, and trading among their participants is intense, often successful, but also frequently highly problematic.

Using the material culture of physicists (while being fully cognizant of the limitations of this metaphor), it seems obvious that our industry is building equipment for a panoply of uses, ranging from banks to research laboratories. Maybe less obvious, each piece of equipment is the vessel for an ongoing behavioral experiment in which the quality of the building plays an obscure role. This obscurity is the cause of much discussion and disagreement in the trading zones. In fact, success itself is a volatile subject: What may be a financially successful building may also be a disaster to work in, and vice versa.

Architects, the most likely participants to talk about their theories, are often blamed by others in the trading zones for being just that: too theoretical. But it is also clear that all the players have their own theories, from economic models to assumptions about the behavior of structural systems. All these various theories form a cluster of behaviors and attitudes that subtly infiltrate the buildings that are constructed. If, then, every completed building is an experiment from a productivity perspective, it is unfortunate that very few post-occupancy studies are done to measure that success or failure. This clearly does not bode well for the arguments made in the trading zones on the relative merit of one building type over another.

From an equipment perspective, the "banging on the bench," as Galison calls it—or the use of the building, in our case—is equally obscure. The distance between the experimenter using a microscope and the data she produces on a physical phenomenon is much narrower than the distance between a CEO and the measurable benefits of the corner window in his office, which presumably helps him to visualize the future of the company. Again, the building industry is placed at a disadvantage by operating in a much more speculative environment. Are there ways to shorten these distances and dispel the obscurity? Most likely. The emergence of the study of buildings in use, and the context in which they came about, leading to evidence-based design are signs of a new era, one that is now clearly more scientific, demanding a better understanding of the performance of buildings as instruments to enhance our productivity and satisfaction.

Joe Powell's book is one of many steps that are being taken toward that end at the RBI, and it recognizes that, just as in groundbreaking science, making buildings is a competitive business, and the success of the 1.5 million companies involved is highly dependent on circumstances outside their control. Happily, and luckily, we can do something to reduce that dependence. This book has the added advantage of being part of a university business, in which unblinking research provides the fundamental basis for its own success as a trading zone. It is also clear that we have our job cut out for ourselves, since while the building industry is known for its willingness to act, it is not yet noted for its willingness to reflect.

Lars Lerup
Dean
Rice University School of Architecture

DEDICATION

For Emma Kate Powell
Who animates all of my endeavors

ACKNOWLEDGMENTS

This book is the result of a team effort. All inaccuracies, confusing ideas, and uninspired prose were supplied by me. On the other hand, intelligent, relevant work was provided by Susan Garza, Mitchell Shields, and Jennifer Mashburn. Special recognition goes to Dean Lars Lerup for conceiving of and founding the Rice University Building Institute. And I'm particularly grateful for the 30 years of insightful, powerful wisdom shared by Ray McLaughlin and Richard Ashley. Hazardous duty pay should be provided to Helen Castle for not acting on her natural impulse to have me expelled from the planet.

Introduction

SHAPING THE BUILT ENVIRONMENT

One and a half million individual companies design, engineer, and construct the built environment in the United States, Britain, Canada, and Australia. What's the single most compelling concern of the owners and managers of these firms? Interestingly enough, that's exactly what we've been asking them for the last three years. The answer: "Competitiveness."

What can we learn about the constant vibration of complex interactions that exist between a firm, its customers, and its competitors? Is it possible to identify the performance characteristics that most accurately determine which firms will achieve market prominence? How will future companies design and execute their competitive strategies?

Firms that function in the AEC industry form highly interactive project teams that work closely together to design and build its buildings. But before project teams are formed, before contracts are signed, before anyone sharpens the lead in their mechanical pencils, an intensely competitive set of events takes place. It's obvious to anyone who's played the game that some companies function in this arena with forceful energy while others struggle to pick up the leftover crumbs.

Some firms compete locally, some compete regionally, some compete nationally and internationally; but interestingly enough, the keys to competitiveness are the same. The Rice University Building Institute has identified 12 areas of competitive focus that will drive the 21st century's most successful AEC firms, regardless of profession, size or location.

Since the subject is how to win more new projects, aren't we talking about marketing? Aren't we talking about sales? Marketing and sales are obviously important, but that's not what this book is about; it's about creating a competitive company. The world's most effective marketing practices will not long elevate a company that lacks "competitiveness."

Research

The material in this book is based on a three-year study executed by Joe M. Powell and the Rice University Building Institute. The Institute set out to define the specific performance characteristics that most powerfully impact competitiveness in the AEC industry.

This undertaking was made somewhat more complex by the fact that the process of shaping the built environment involves a diverse collection of professionals and disciplines. We were not interested in how any one of these groups functions in a vacuum, because they acquire their power and utility to society only when they effectively and efficiently interact. Therefore, we have chosen to investigate the AEC industry as a whole.

Approach
1. Define "market prominence."
2. Identify firms in all AEC disciplines that have achieved it.
3. Investigate how they did it.
4. Define what performance characteristics will be required of the next generation to acquire a position of market prominence.

Surveys
The Institute surveyed 28 of America's leading graduate schools of business.

Interviews
The Institute interviewed 164 company principals from the fields of architecture, design, engineering, general contracting, specialty contracting, project management, and program management. We also involved representatives from the major corporations that own and manage large real estate and facility portfolios.

Focus groups
The Institute invited 102 industry leaders to participate in a series of on-campus focus groups. These groups were designed to be small and highly interdisciplinary. Their interaction was fascinating and instructive.

The Rice University Building Institute

An interdisciplinary collaboration of industry, community, and academic leaders
Rice University has a distinguished track record in establishing and operating interdisciplinary collaborations in teaching and research. The university currently is home to over 40 research centers, institutes, and consortia.

The modern process of shaping the built environment has become so complex that needed improvements and innovations lag far behind the growing challenges. Clearly, innovative approaches will be required to re-integrate what has become a fragmented process. This won't happen without the creative interaction of our most respected leaders from industry and academia.

The Rice University Building Institute provides the requisite forum in which this interdisciplinary search for innovation comes to life.

The Institute has four operational centers:

- **Research:** Defining new insights
- **Executive education:** Improving performance
- **Publishing:** Disseminating the latest findings
- **Symposia**: Hosting interactive exploration

Institute stakeholders
- **Owner/user groups:** The corporations that own and use large real estate portfolios
- **AEC industry**: Architectural profession/engineering profession/construction industry
- **Building material manufacturers**
- **Financial industry:** Those who finance and insure the building development process
- **Professional advisors:** Real estate brokers and consultants/real estate attorneys/real estate accountants
- **Developers:** Investment builders and merchant builders

How to use this book

You will find five types of information intermingled herein.

(1) Questions

?

Research is about answering questions; here are the ones we asked.

(2) Descriptive text

This is what we learned about each of the 12 performance characteristics found within companies that achieve market prominence.

(3) Book recommendations

You will notice that each characteristic describes an element of company performance that has been studied for years by an entertaining mix of management experts and jesters. Our job here is not to present another explanation of these basic business functions, but rather to explain how you can use them to improve competitiveness. Therefore, for those interested in a more in-depth explanation, we recommend books that have achieved general recognition for their relevance.

(4) Industry stories

We invited industry leaders from the fields of architecture, engineering, and construction to provide personal stories illustrating their approaches.

(5) Industry leader comments

Research is useless unless it finds its way, sooner or later, to utility among practitioners. Therefore, we have invited a diverse group of prominent professionals to comment on our work. Some of their opinions were non-libelous and have been integrated herein.

Some companies in the AEC industry are able to acquire and maintain a perennially imposing presence in the marketplace. How do they do it? More importantly, how will the next generation do it?

EXECUTIVE OVERVIEW

It's a phenomenon with which we are all familiar. Some companies in the AEC industry are able to acquire and maintain a perennially imposing presence in the marketplace. How do they do it? More importantly, how will the next generation do it?

> **AEC industry**
> Architects, designers, engineers, general contractors, specialty contractors, major subcontractors, project managers, program managers, construction managers

Conventional wisdom tells us that architects compete against other architects, engineers against other engineers, general contractors against other general contractors, and so on. It makes sense, then, that certain competitive strategies could exist that are effective in one discipline and not in another.

We found no such thing. Selection processes and fee negotiations vary somewhat, but the basic performance characteristics that propel one firm past its competitors are the same.

What about location? Do the designers and builders in Britain face competitive forces different from those in Canada, Australia, and the US? First, we found more companies functioning successfully outside their home country than ever before. And even though we have witnessed an entertaining collection of terminologies that don't always match, the competitive forces around the world are identical.

We found the same rule of invariance with respect to company size. We were gratified to find competent, profitable, well-regarded firms of all sizes in every category. And even though size impacts the type of projects these firms pursue, the performance characteristics that matter work equally well in each.

We did uncover fascinating variations in culture, however. Behaviors that are prized in one profession will get you ridiculed in another. While these divergent experts work together everyday to solve building problems and typically have great respect for one another, they are fiercely proud of their own professional identity. The engineers don't want to act like architects, the architects don't want to act like engineers, and the builders don't want to act like either one.

> **I**'m not so concerned that we don't know all the answers. What worries me is that we don't know the right questions.
>
> LEON LEDERMAN
> NOBEL PRIZE-WINNING PHYSICIST

The important questions

Research is about answering questions. In fact, asking the right questions is the single most important part of the process. The reason that so many research findings end up being published in the *Journal of Irrelevant Findings* is that the researchers begin by asking the same questions they always ask, looking in the same places for answers that they always look, and, not surprisingly, ending up with the same tired, old ideas.

So, what are the important questions?

1. Since we're interested in the concept of market prominence, how shall we define it?
2. How did the current generation of AEC market leaders acquire their lofty positions?
3. What will be required of the next generation who aspire to market leadership?

(1) How shall we define market prominence?

Achieving a position of market prominence is undoubtedly the holy grail of business performance. It's actually a more powerful experience than managing to dominate a market for a period of time. **Market dominance** is typically an exciting but brief phenomenon and usually demands a hefty modicum of chance. **Market prominence**, on the other hand, connotes an imposing position that endures. It isn't luck. It can't be achieved by commercializing the latest fad. You don't get there by being in the right place at the right time. It belongs solely to those companies who consistently and repetitively do the right things.

While it's tempting to define the term "market prominence" quantitatively, our researchers chose to utilize qualitative measures instead. We feel that the power of this term lives not on a spreadsheet, but in the hearts and minds of those entities that define a market (customers, competitors, employees). It's actually a simple matter. Take any specific market segment, survey the major buyers, the major participating firms, the appropriate professional associations, and you will quickly identify those companies considered to have achieved "market prominence." In fact, in your arena, you already know who these firms are. Here are the characteristics we measured to determine market leaders:

Name recognition
Ask major participants in any market segment and they tend to repetitively mention the same names.

Expertise recognition
Market leaders are successful at attaching, in the minds of major buyers, qualities associated with advanced expertise and performance, whether or not that advanced expertise actually exists.

Premium pricing
Buyers honestly believe that they're going to get more value from market leaders and are typically willing to pay for it.

Preferential competitive treatment
In a competitive sales arena, clients often let market leaders skip the "qualifications" stage of the competition and go directly to the final sales presentation.

Prestige among employees
Survey professionals in any defined market area and ask them about the best places to work. You will find a high correlation between market prominence and desirable work environments.

(2) What performance characteristics will determine the next generation of market leaders?

Everyone knows predicting the future, no matter how much data you have, is a dicey proposition. People substantially smarter than us have perished in the attempt. Nonetheless, this project and this book are largely useless unless we can provide our readers with some insight into how future market leaders are likely to acquire prominence.

Conventional wisdom tells us that yesterday's market leaders achieved their positions by consistently applying three practices: aggressive personal networking, effective competitive sales, and the capacity to perform as promised.

Today's game is more complex. We believe there are 12 performance characteristics, each requiring its own executable strategy, that will be essential to consistently outperforming the competition. You will notice that we do not discuss three issues: financial stability, ethical behavior, and competent professional performance. These are assumed.

Here are the 12 performance practices that will drive tomorrow's most competitive firms:

Today's game is more complex. We believe there are 12 performance characteristics, each requiring its own executable strategy that will be essential to consistently outperforming the competition.

1 Vision
Vision-driven companies attract better employees, function with more purpose, and are more adept at holding the attention of high-value clients. Unfortunately, this requires that someone actually have a perceptive vision and the capacity to rally others to it. Not an easy task.

2 Values
In today's greedy, short-cut driven culture, values have never been more important or precious. But this study is not about finding companies that do the right thing, it's about defining those

characteristics that contribute to competitiveness. Interestingly enough, companies that genuinely nurture a culture of values, over time, are more effective in the marketplace. Even though we witnessed several brave attempts, this is one of those characteristics that is impossible to fake.

3 Competitive focus

It's confusing out there. Attractive distractions are everywhere and strategic focus has never been more difficult to maintain. Tomorrow's most competitive companies will know their mission and stick to it. That's not to say they will ignore genuine opportunity, but they won't chase off in various directions capriciously.

4 Category ownership

It's common knowledge that dominating a market category, even temporarily, can be very profitable. But what if all of the categories in which you compete already have imposing market leaders? Future competitors will know how to invent a new category and declare ownership. Sounds audacious, but it's perfectly achievable and we'll explain how.

5 Persistent branding

Branding is a phenomenon that has been studied in the business community for many years. Even though its value has been clearly established, few companies in the AEC community effectively exploit its power. We found several examples of companies that do it intuitively, but tomorrow's leaders will execute a specific branding strategy.

6 Marketing breakthroughs

There will be certain projects for which your company is considered a "favorite" and others for which your company would be considered a "long shot." The acquisition of market prominence requires the orchestration of "breakthroughs," winning projects for which you are not the most qualified competitor. Tomorrow's market leaders must develop the capacity to walk into a room filled with more experienced and credentialed competitors and walk out with the project. Companies who can't pull this off will be destined to an eternity of similarly repetitive projects from similarly repetitive buyers while generating uninterestingly bland margins.

Today's AEC market leaders are adept at discussing design and building strategies. Tomorrow's leaders will be just as comfortable discussing their clients' business strategies. Owners spend enormous energy developing marketing strategies, production strategies, capital management strategies, and human resource strategies, and they will expect AEC experts to explain to them how the built environment can be used as an asset in executing these identified goals.

7 Competitive intelligence

We found that most companies have only a very intuitive and inadequate view of the competitive landscape. Not surprisingly, the next generation of successful competitors will have more accurate information about the specific rules of engagement.

8 A competitive culture

Within many companies, the responsibility for tracking the competition, looking for new opportunities, and nurturing a competitive advantage is assigned to a relatively small group. Often that group is the marketing and sales people. The new market-leading companies won't work that way. Everyone in the venture must be cognizant of the competitive landscape and will be rewarded for the constant mining of new opportunities.

9 Customer intimacy

These days, most successful companies do fairly well at understanding the stated needs of their customers. And they have a passable understanding of how to engage them. In a highly competitive environment, however, new market leaders must find a way to know more, to uncover high-value customer needs that are not common knowledge. Every major buyer has motivators that remain unarticulated and can't be captured by conventional market research.

10 High-impact people

Normal, well-adjusted people often make for pleasant dinner companions but seldom consistently outperform the competition. New market leaders will purposefully identify, acquire, and manage high-impact people who are worth more than their pay. This kind of thing is harder to do than it sounds, but it is an invaluable component in achieving breakthroughs and outperforming the competition.

11 A culture of obsessive improvement

Self-congratulatory ventures don't survive long. The most competitive companies seem to operate within a culture of obsessive improvement, and they require as much from employees at every level. This is the kind of thing that's fun to talk about, so that's what most companies do. Few, however, are actually willing to pay the price. For those who do, the rewards can be substantial. We found that the most successful ventures institutionalize the process.

12 New strategic alliances

In the next ten years, differentiation will be hard to achieve within the boundaries of any individual discipline. Project delivery systems are broken and everyone knows it. Buildings take too long to design and build, cost too much, and have difficulty responding to the ever-changing business demands of the owners. Desperately needed improvements simply can't be achieved by any one of the individual disciplines involved. As horrible as it sounds, we're actually going to have to talk to each other. The next generation of leaders must lead the effort to exploit the power of interdisciplinary collaboration.

Today's AEC market leaders are adept at discussing design and building strategies. Tomorrow's leaders will be just as comfortable discussing their clients' business strategies.

Vision

Several years ago, an international executive search firm surveyed 1500 senior corporate leaders from 20 countries.

The question: "When selecting a CEO, what performance characteristics do you consider to be the most valuable?"

The answer: "A clear vision and the ability to rally people to it."

Chapter 1

VISION

What constitutes an effective vision statement?

How are effective vision statements developed?

What are the typical impediments to becoming vision-driven?

How does being vision-driven translate to elevated competitiveness?

Vision: How Leaders Develop It, Share It, and Sustain It
Joseph V. Quigley
McGraw Hill, 1993

Several years ago, an international executive search firm surveyed 1500 senior corporate leaders from 20 countries.

The question: "When selecting a CEO, what performance characteristics do you consider to be the most valuable?"

The answer: "A clear vision and the ability to rally people to it."

The subject of corporate vision has gotten plenty of attention over the past 20 years and as is typically the case, it has spawned a cottage industry of consultants, authors, facilitators, and tricksters. These days, most companies, even small ones, have written vision statements, values statements, and strategic plans. Obviously, some of these documents serve as the basis for energizing and focusing superior performance, while others are nothing more than frilly window dressing at which employees snicker when the boss's back is turned.

What constitutes an effective vision statement?

The next generation of market leaders will use motivational vision statements to add bandwidth to all of their marketing and communication programs. We identified five characteristics shared by the more successful ones.

1 It addresses the future
A good vision statement is about the future. Properly formulated, it will become your group's most powerful organizing force. An

effective vision is an emotionally compelling portrait of what you intend to become. It acknowledges the past, is cognizant of the present, and describes a passage to a more fulfilling tomorrow.

> Our vision is simply to work with people, through architecture, to evoke the world to which we all aspire.
>
> CHRISTOPHER RATCLIFF, PRESIDENT AND CEO
> RATCLIFF ARCHITECTS
> SAN FRANCISCO, CALIFORNIA

2 It is emotionally compelling
Business is not only a game of the intellect. Even though we spend enormous energy discussing and admiring our intellectual prowess, most business behavior has an overwhelmingly emotional component. This perhaps is due to the large number of human beings involved. Here's the good news about studying the human emotional condition: it isn't changing. Business environments change, practices change, customers change, employees change, technology changes, but there has been no measurable evolution of basic human emotional drivers in the last two million years. OK then. Here's the "so what":

Companies that emotionally engage their stakeholders can simply ask more of them—more of their customers, more of their employees, more of their consultants.

> Our design consultancy is dedicated to the reintegration of architecture, landscape, and the civic arts. We are committed to a significant restoration of the storytelling qualities of architecture.
>
> ERIC KUHNE, MANAGING DIRECTOR
> ERIC R. KUHNE & ASSOCIATES
> LONDON, ENGLAND

Effective leaders have always known how to exploit this reality. An emotionally compelling vision will be one of your most powerful assets for motivating employees beyond simply doing their jobs. And it may serve as an unusually powerful point of differentiation between you and your competitors.

Emotional Intelligence: Why It Can Matter More Than IQ
Daniel Goleman
Bantam, 1997

3 It is reality based

Vision statements are intended to motivate. There is nothing quite so pitiful as a group of people who pretend they are committed to a set of goals which they know are clearly out of reach. Your vision must be aggressive yet achievable. So how do you find the sweet spot? It's clearly a matter of negotiation and one of the most important elements of your vision development process.

4 It describes a meaningful/deeper purpose

We found several companies led by owners who believe the free enterprise system supplies all of the meaning any company needs. "We're here to make money for our investors and ourselves. What else do you need to know?"

In the 1960s, when General Motors began to feel the initial indicators of what may have become its inexorable decline, its board decided that they needed a new CEO to provide a spark of creativity and energy. On his first day, James Roche held a press conference where a reporter asked, "How does it feel to be responsible for making more cars that any other company in the world?" Mr. Roche's response: "I wasn't hired to make cars. I was hired to make money." Wow. That's a pretty clear sign of the vision he intended to implement.

Here's how it typically goes: "We intend to provide our investors with a superior return, our customers with superior service, our employees with a superior place to work, while being good corporate citizens." How's that for a statement that's cool-gray, vapid, and powerless?

There are many people who believe our lives are not about a search for happiness, but a search for meaning. If that's the case, there will be many rewards for any management team that can make a clear connection between the work of their company and the personal meaningfulness for their workers.

Seventy years ago, the founders of Perkins & Will described a vision for their fledgling architectural practice. "We intend to produce ideas and buildings that honor the broader goals of society." According to Greg Hughes, Principal, "Every associate here knows that our work is not about us. It's about how we impact society."

Man's Search for Meaning
Victor Frankl
Beacon Press, 1954

5 It permeates everyday work

We found several examples of companies run by executives who were personally guided by a clear vision. Unfortunately, they were the only ones who knew about it. So is it better to have no vision whatsoever, or one locked between the ears of people in senior management? Who cares? The results are the same.

Meaningful visions that actually impact operations aren't hard to identify. All you have to do is spend a little time at the company. It will be readily apparent that everyone is aligned on a few sali-

Permeate Everyday Work
How do the most competitive companies permeate everyday work with their vision?
The bad news: The mouth will be your least effective instrument in communicating your company's vision.

ent concepts. That doesn't mean that all leaders are the same or that all employees are the same. In fact, a great deal of personal diversity can and probably should be present, but everyone needs to know why they're there and what role they play. Company visions that don't live in the hearts and minds of all key players simply can't materially impact the way things get done.

So, how do most competitive companies achieve this?

First, the bad news.

Your mouth will be your least effective instrument in communicating your company's vision.

The most effective practices that we observed were very simple. First you tell them about it (not very difficult) and then you live it (harder).

At Linbeck in Houston, every new hire spends four hours with Chairman Leo Linbeck, III, and President Chuck Greco. According to Mr. Greco, "This kind of personal attention is expensive for the firm, but it's where we begin the process of drilling our company vision and values."

Failures at Creating a Vision
Failed attempts at designing and implementing corporate visions are numerous and everywhere. Why do so many good companies spend so much time and money on this topic while producing such paltry results?

> We teach a wide curriculum at Jacobs College; but by far, the most valuable thing we teach is our company DNA.
>
> ROBERT M. CLEMENTS, SENIOR VICE PRESIDENT
> JACOBS ENGINEERING GROUP INC.
> PASADENA, CALIFORNIA

How are effective vision statements developed?

The most effective vision development process for any particular venture clearly depends on the key relationships that exist between company owners, senior management, and key employees. In this discussion, you'll notice the repetitive use of the term "key employees". No sizeable company ever gets everyone on board at the same time, but the good news is that it's not necessary. Not everyone in your company contributes equally, so pick the most valuable players and make certain you've acquired their commitment.

> It seems quite impossible to communicate my vision to all of my 120 associates individually, so I find myself spending time with those who demand it.
>
> MASSIMILIANO FUKASAS, OWNER
> MASSIMILIANO FUKSAS, ARCHITETTO
> ROME

Regardless of how a vision statement is developed and communicated, one reality is perfectly clear. All owners, key managers, and key employees must absolutely buy in. The problem is: How can you tell who is committed and who isn't? Smaller firms regularly make this value judgment intuitively, which can work perfectly well. Larger firms, however, need evaluation techniques that are more formalized. The cohesive effort required to achieve market leadership can absolutely be dismantled by a few dissidents who aren't devoted to the quest.

We observed successful vision statements being developed in a variety of ways, but three specific techniques were common:

1 Announce

If the owners founded the company with a clear vision and feel strongly about it, the most efficient process is to simply announce it and live it. People who buy in can stay and those that don't should work elsewhere. Once again, letting those who don't buy in hang around is a big mistake. Get them on board or get them out of the way.

> **O**ur design philosophy and company vision were inspired by one individual. The process of translating that individual's vision into a collective one comes from breathing the same air, working closely together, and constant communication. It is not the result of slogans, posters, or vision programs.
>
> RON KEENBERG, CHIEF ARCHITECT
> IKOY
> OTTAWA, CANADA

2 Consult

A consulting strategy makes sense when senior management makes it clear that they are ultimately responsible for writing the vision statement, but input from key employees is important. This requires a systematic listening process but typically takes less time than a more collaborative one.

3 Collaborate

As previously discussed, most of us have a compelling emotional need to feel that our work has meaning and that it fits into a larger purpose. Therefore, many highly competitive firms utilize a process that requires all relevant stakeholders to join together to co-create the firm's vision. While this technique can offer the greatest rewards, it is also home to the most pitfalls.

> **O**ur company is 143 years old and the substance of our vision has never changed. Every five years, however, we take our entire company through a formal visioning process, not because we're interested in rethinking our vision, but because the process is invaluable in creating emotional buy-in on the part of our key people. Our 143-year-old vision becomes their vision.
>
> CHRIS PECK, VICE PRESIDENT
> MCCARTHY BUILDING COMPANIES
> ST. LOUIS, MISSOURI

Even though individuals will always have somewhat divergent motivators, all key people in a vision-driven company share an overarching sense of purpose.

The greatest downside to the collaborative process occurs when senior management is not prepared to listen and key employees are not prepared to meaningfully contribute. Chaos and polemics typically follow. So what's so difficult about establishing an honest idea exchange between a firm's owners and its most valuable employees? It seems like this should just naturally occur. It doesn't.

We witnessed several senior management teams produce what they considered to be an honest effort to actually connect to their people only to have the exercise end with frustration or, even worse, palpable indifference.

Honest human connection only happens in an atmosphere of intellectual and emotional safety,

a condition often missing from these kinds of proceedings. Why?

The Fifth Discipline Fieldbook
Peter M. Senge *et al.*
Nicholas Brealey Publishing, 1994

Perhaps it's because we found the ranks of senior management were often colonized by a plethora of gifted talkers—charming people who can work a room, tell a good story, and generally entertain. While this performance characteristic is highly prized by engineers, builders, and architects alike, it's useless when the goal is to stimulate powerful sharing. For that, you need active listeners—people who are genuinely interested in the ideas of others, otherwise known as consensus builders.

All previous options are preferable to an insincere co-creation attempt.

Developing Our Company Vision
Jim Jonassen/Scott Wyatt
Managing Principals
NBBJ
Seattle, Washington

For NBBJ, creating a compelling vision that people rally around meant cracking the process of creating an architectural firm monograph wide open and driving a series of endeavors through the firm at all levels—firm, studio, and individual. We started by engaging the firm's 67 principals. We asked the question: "When it's as good as it gets, what are we doing?"

We held workshops, conducted interviews, and collected stories about best practices. We were intentionally provocative. The core communications team on the project set a metric of success; we'd know we were succeeding if we managed to "draw fire" from more than half the participants.

To kick off a lively debate, we created a working theme, *Fire the Architects*, and mocked up a book jacket with that title. Needless to say, it worked. People had ideas, opinions, perspectives, and anecdotes to share. The provocation helped people become deeply engaged in the process. These initiatives helped capture the firm in book form and, on a deeper level, they were also crafting and shaping the courageous culture itself, enriching the firm's mission and strategic focus through inclusion and dialogue.

After a series of explorations, we settled on the title, *Change Design*. The double entendre served a dual purpose—it spoke to our clients' need to constantly address unprecedented rates of change in their business arenas. It also spoke to the need for our own profession to change the practice itself by linking services and design solutions more closely to our own clients' enterprises.

In designing the design firm where creative people do their best work, the *Change Design* movement illustrates NBBJ's vision to balance freedom and exploration with structure, hierarchy and standardization.

The lessons learned we came away with were numerous. Two take precedence.

The first is that process is equally as important as outcome.

The second is that when you are designing an organization for creative growth and competitive advantage, it often means unhitching from the single guru figure of the managing board and allowing the entire group to shape the firm, from brainstorming to implementation. We structured a cellular system whose people are interdependent. When the culture creates its own roadmap, the organization has more of an unlimited growth potential.

What are the typical impediments to becoming vision-driven?

Failed attempts at designing and implementing corporate visions are numerous and widespread. Why do so many good companies spend so much time and money on this topic while producing such paltry results? Interestingly enough, we found that most failures occur for very predictable reasons.

First, let's discuss what's not causing the problem. The desire to be a vision-driven company was almost universal. Next, every company we investigated was perfectly competent at generating a vision. Some were more inventive than others, but coming up with an acceptable vision was not a problem. Although most companies in the AEC industry do not hire consultants to help generate their visions, we observed a clear willingness to spend time and money to get it done. Let's recap: There is a general desire to become vision-driven, most groups do perfectly well at writing it, and almost everyone is willing to assign resources to the undertaking. So why is there such a high failure rate? Why do so few companies actually become vision-driven?

Repetitive failure occurs because a virtual firewall develops between those who create the vision and those responsible for executing it.

As we've repeatedly emphasized: the vision, no matter how well-crafted, is useless unless it lives in the hearts and minds of all key employees. Here are some typical problem areas:

Failures at creating a vision

Failed attempts at designing and implementing corporate visions are numerous and everywhere. Why do so many good companies spend so much time and money on this topic while producing such paltry results?

1 Communication
Some companies simply don't communicate their visions repeatedly. It's not good enough to bring it up now and then. Vision-driven companies talk about it constantly.

2 Personal
Company visions lack valence when they don't fire the imagination of key workers. A vision is powerless until it becomes a personal imperative.

3 Measurable
Any management program whose performance isn't measured will remain useful for water cooler talk and little else.

Several years ago, I accompanied a friend who dropped his five-year-old daughter off at day school. He said, "If you've got a few minutes, let's go inside. There's an important management lesson I want you to witness." Parents are allowed to observe from a special enclosure where the kids were not aware of their presence.

When we arrived, some of the 12 four to five-year-old students were milling around the room basically doing whatever they cared to. Some were talking, some reading, some playing, some staring out the window. When it was time to start class, the teacher took a seat at the head of the room and began singing the "Good morning, it's time to start" song. Hearing this, several of the children immediately took their seats and began paying attention while others went right ahead with their self-selected activities. This went on for a while and my friend said to me, "Notice that she never scolds them." After several more minutes, 2 or 3 additional children began paying attention and took their seats. Shortly, a few more came to order. Meanwhile, the teacher is still singing. Finally, the last group of stragglers took heed and joined the class. The teacher never raised her voice or threatened the kids, she just kept singing the song.

There's something to be learned here about communicating your group's vision and values. Don't expect everyone to buy in at the same time. Just keep singing the song.

Obviously, there's a time and patience limit with the strategy. After a reasonable period, identify the disruptors and get rid of them.

4 Rewarded
Specific rewards must be given to individuals for advancing the group vision.

What about laggers?
We identified another interesting phenomenon relative to the task of integrating a coherent vision. No matter how well you communicate your vision, no matter how well you measure and reward those who actively help achieve it, there will be those who simply don't buy in. On the surface, this seems innocuous enough. The problem is that these laggers can wield enormous influence.

The question quickly becomes: Who is with us and who isn't? Small firms deal with this intuitively, which is usually fine. Large ventures need more formalized techniques to identify players who are not actively advancing the group's vision.

We identified six typical employee responses to any new vision initiative and offer our thoughts about how senior management should deal with each.

Ignore
New initiatives are inherently scary and there will be those that simply hope it goes away. These people will act as if nothing new or important is happening.

1 Ignore
New initiatives are inherently scary and there will be those that simply hope it goes away. These people will act as if nothing new or important is happening.

Response: Leadership must quickly identify the laggers and either bring them on board or get them out of the way.

2 Sabotage
Fear often causes some people to feel uncomfortable with the honest expression of opposition and they may choose to work behind the scenes to derail the vision. While being overtly supportive, they will secretly attempt to damage the effort. This response is unhealthy for the company as well as for the individual.

Response: These people must be identified and quickly put on a short leash. Try getting them involved. If that is not immediately successful, remove them. It won't take many of these people to keep a coherent vision from taking hold.

3 Overt opposition

There will always be perfectly rational reasons for an individual not to support any new vision. While sabotage is an unhealthy response to a proposed change, overt opposition is the honest expression of disapproval.

Response: Leadership should deal with these individuals with dignity and respect. Sell them on the reasons for the new vision, involve them in communicating it, and reward them for their support. Monitor them closely. If rapid 'buy-in' is not achieved, they must be replaced.

4 Wait and see

Change is scary, but then again, it might improve things. We often heard: "I'm not crazy about having to think differently about this company or my work, and I'm not certain whether this new thing is for real or just another passing management fad. I think I'll just wait and see how things play out."

Response: As with the people in categories 1, 2, and 3, leaders must quickly identify the "wait and see" types and either get them on board or remove them.

5 Go along

If leadership support for the new vision statement seems overwhelming and its value to the company has been made clear, some will just go along with it. They won't oppose it, but they will not go out of their way to support it.

Response: While these people will seem initially tolerable, they won't help you get where you need to be.

6 Active support

People in this category were typically emotionally invested in the development of the new vision statement in the first place and will actively work to make it real.

Response: These people should be identified and publicly recognized. Reward them with key assignments.

How does being a 'vision-driven' company translate to increased competitiveness?

There's a difference between a company that has a vision statement and a company that is vision-driven. Actually, vision statements aren't very hard to write. In fact, any bumptious consultant smarter than the tassels on his Gucci loafers can whip one out for you in no time. Not inexpensively, but quickly.

Before we can discuss why vision-driven companies usually outperform those that aren't, we must agree on a working definition of what it means to be vision-driven. Our research partners settled on this:

A vision-driven venture is one in which all goals, missions, and strategies are animated by a meaningful sense of shared purpose.

Project teams
In the design and building industries, however, the predominant organizational element is the **project team**. It therefore makes sense that it is just as important that each substantial project team be vision-driven. Team members find it much easier to align daily activities with stated project goals when an overarching sense of common destiny is present.

McLachlan Lister Pty Ltd Project & Strategy Advisors in Sydney assembles interdisciplinary teams of designers and builders for large infrastructure and commercial projects in Australia. Describing themselves as "ego wranglers," CEO Leslie Butterfield explains that their goal of creating a high-performance project delivery team is heavily dependent on selecting individual members who are effective communicators and willing to commit to a coherent team vision. Says Ms. Butterfield, "We routinely look beneath individual company credentials and select specific professionals who, we believe, are likely to function effectively in a seamless project team."

So when you walk into the reception room of any given company, how do you tell to what extent the venture is vision-driven?

We quickly discovered that it is useless to simply ask people in management. Since having a corporate vision is fashionable these days, all companies claim to have one. Further, it is very unusual to find individual partners or principles who aren't motivated by a personal vision of some sort. But, that, in fact, is not the issue. We're trying to define whether or not the company itself is vision-driven. Here are the key indications we used:

1 Shared
Even though individuals will always have somewhat divergent motivators, all key people in a vision-driven company share an overarching sense of purpose.

2 Conscious
When you ask about the company's vision, the key players don't have to refer to a poster on the wall or quickly run to the employee manual to look it up. It lives in the forefront of their consciousness and impacts their decisions on a daily basis.

Even though individuals will always have somewhat divergent motivators, all key people in a vision-driven company share an overarching sense of purpose.

3 Measurable
A vision that is charming but philosophical may provide a nice, warm feeling, but is otherwise useless. Vision-driven companies carefully translate their vision into specifically measurable goals, missions, and strategies. And they know exactly where they stand in the process of achieving each.

4 Personal
The group vision genuinely relates to each key individual's aspirations. We find that when an individual's personal vision is in conflict with that of the group, commitment lags, performance lags, and the group vision becomes irrelevant.

When I see vision statements displayed on t-shirts, posters, or cardboard conference table signs, I become instantly concerned that these people have missed the point.

If your vision statement was written by the PR department, a consultant, or your marketing team, you might as well have hired a greeting card company.

Beautifully displayed, vapid vision statements have become a staple of enterprises in decline.

Human beings, all of us, are predominately emotionally-driven creatures and the most competitive companies find effective techniques for engaging all of their stakeholders emotionally. An engaging vision is simply one powerful way of achieving that.

Here are the measurable benefits we found:

1 Employees
You will attract better employees.

2 Marketing
Marketing is about getting your face in front of the face of a potential buyer. It's crowded and confusing out there. A vision-driven marketing strategy often presents a more cohesive face to your targeted clients. Vivian Manasc, a principal with Manasc Architects and 2007 president of the Royal Architecture Institute of Canada, says that "Our vision is not just about what we want to become. It's about which clients we want to work with."

3 Sales
Competitive selling is about quickly differentiating your company. We've seen this achieved in several ways, but owners clearly respond to companies that seem to be driven by overarching concerns that transcend their work on any one project.

> We don't have a single piece of marketing material—not a brochure, not a pamphlet, no sales force, no advertising. We are only hired for the vision we have.
>
> ERIC KUHNE
> ERIC KUHNE & ASSOCIATES
> LONDON

4 Project teams

Our work is done by temporary, interdisciplinary organizations often referred to as **project teams**. It makes no sense for an individual company to have a vision for their work that never gets inserted into the combined consciousness of the major players on the project teams. And yet, we observed this over and over again. Even though teams are typically created to accomplish only one specific building project, a focused vision lives at the heart of those that outperform the norm.

> We regularly evaluate all of our project teams and purposefully judge them, not on the performance of any individual company, but on the work value of the team as a whole.
>
> DAVID CHAMBERS, DIRECTOR, PLANNING ARCHITECTURE DESIGN
> SUTTER HEALTH SYSTEM
> SACRAMENTO, CALIFORNIA

5 Development process

We observed that the vision development process itself can serve to powerfully increase competitiveness. Several market leaders involved not only their key employees, but their key consultants and targeted customers. Several others involved the heads of local professional associations such as the American Institute of Architects (AIA) or Association of General Contractors (AGC) to provide an industry perspective. The strategy of using an extended group beyond your company to develop vision and mission statements was particularly powerful.

High-performance, competitive companies typically operate with an underlayment that unifies their divergent experience and expertise into one cohesive, meaningful quest. That substrate usually turns out to be a powerful vision.

Values
Values are an underlying set of adopted personal assumptions that guide every individual's judgment and behavior.

Chapter 2

VALUES

Do values actually make a difference?

How are today's market leaders dealing with values?

How will tomorrow's market leaders utilize values to be more competitive?

Values ... not a new topic. Aristotle talked at length about them, as did Confucius. Since then, every saint, dictator, evangelist, politician, motivational speaker, and magazine salesman has invoked the chimerical power of values to bend the masses to their purposes. Today, of course, no self-respecting corporation would be without some sort of statement delineating their high-minded aspirations to be the best possible citizen of the universe, to always do the right thing, and to respect the rights of all the Earth's creatures.

Our concern is: Besides all the warm, fuzzy feelings that emanate from a discussion of values, do they actually make a difference? Do they impact individual behavior? Do they impact team performance? And, of course, the question in which we are most interested: Do they impact competitiveness?

The central task for any company in the AEC industry is to create an environment in which independently competent professionals can collaborate for mutual achievement. Undoubtedly, the two most powerful organizing elements at our disposal are:

- Vision ... where are we going?
- Values ... what rules govern our journey?

Since we need a working definition, let's use this one: **Values are an underlying set of adopted personal assumptions that guide every individual's judgment and behavior.**

Just for fun, we asked our friends in psychology and sociology where human values come from, but the answer was so complicated and uncertain that, and for our purposes, we chose to ignore the issue entirely.

How are today's market leaders dealing with values?

All people and all companies have a functioning value system and we found no current practitioners who aren't actively dealing with the issue. Some develop, teach, and discuss values quite purposefully, while others just let them happen as a result of normal, everyday behavior. But since it is now fashionable to openly discuss values, we found that that's exactly what many companies do. They just talk. In any case, the topic of values is now front and center for most designers and builders.

Leading With Values
Kim S. Cameron
Cambridge University Press, 2006

> We typically don't ask our designers and builders about values because the best ones naturally integrate the material into their presentations. We also find that the consultants we hire spend just as much time interviewing us about our values as we do them about their's. For us, it's not about designing and building individual healthcare facilities, it's about long-term relationships that must have complementary values at their foundation.
>
> DAVID CHAMBERS, DIRECTOR PLANNING ARCHITECTURE DESIGN
> SUTTER HEALTH NETWORK
> SACRAMENTO, CALIFORNIA

What concerns today's market leaders when they think about value systems? Almost everyone intuits that a strong value system is critical to high-level group performance, but this is one of those topics that is hard to clearly quantify. We are not aware of anyone who has developed a credible quantitative relationship between values and financial outcomes. It appears this is one of those issues that must be approached intuitively, or not at all.

We are particularly concerned that, in many industries, corporate fiscal performance trumps all other concerns and that often management or leadership programs that don't deliver measurable financial results are quickly marginalized by all constituents. We believe it imperative that the topic of corporate values not meet that fate. Even in the absence of generally accepted metrics, values are unquestionably worthy of our attention and research.

So, what is the latest thinking on this mystical subject? An in-depth discussion almost always creates more questions than it answers. Here's the way it typically goes:

Development
How should we develop our system? Should we use a consultant? Does a formal process make sense or should we just let our group values occur naturally? Even though our values have been in place for many years, should we go through some sort of process anyway?

Integration
How do we convince all key employees to adopt them? What integration techniques work and which ones seem trite? Does formal training work? What about the new people we hire? Should they get some sort of intensive course about our values? How do we tell,

What integration techniques work and which ones seem trite, when dealing with values?

before we hire someone new, if the values they already have are in concert with ours? How do we tell which of our employees has bought in and which hasn't? Does it make any difference whether or not everyone in the company adopts our values?

Evaluation
How do we tell if all of this makes any difference? Is it possible to evaluate the impact of our values on company performance? Is it possible to tell which of our employees is actively applying our values to our projects?

Reward
Should we formally recognize people for their adherence to and development of our values? Should we pay them more or create some sort of bonus system?

> When we're putting together a team of architects, engineers, and builders, we never ask about their values because they'll simply tell us what they think we want to hear. During visits to their offices, we routinely ask that the sales people stay in the conference room while our team spends several hours walking around their work space visiting with their people. Our selection process is specifically designed to uncover the group's system of values, which you can't do by interacting with people responsible for marketing.
>
> JERROLD P. LEA, SENIOR VICE PRESIDENT
> HINES
> HOUSTON, TEXAS

Development
The object of the exercise, for any firm of any size or age, is to employ a vigorous system of values as an organizing, motivating, and focusing force. Some firms achieve this without ever mentioning the word. No talk, no posters, no training. They just live it.

Diamond + Schmitt Architects in Toronto is a perfect example. With 14 partners and 120 employees, this is a sizeable venture and even though values are important to them, they feel strongly that

Inside the Box: Leading with Corporate Values to Drive Sustained Business Success
David Cohen
John Wiley & Sons, 2006

the power of their value system must live in the details of their daily execution. According to Don Schmitt, "We are committed to the idea of collegial problem solving. When we do it correctly, it provides all of the interdisciplinary interaction necessary to keep our values alive and well. Everyone's ideas are given respect, innovation and risk are encouraged, but the specific subject of values never comes up."

Actually, for our purposes, the topic of value system development is not a particularly fertile one. Every firm we investigated has one and is committed to it. The more compelling issue is how to integrate long-held values into the fabric of a complex venture.

> I work for an 8 billion dollar company with 48,000 employees and a well-established set of corporate values. What we worry about is how to get the message to the 48,000th employee.
>
> ROBERT M. CLEMENTS, SENIOR VICE PRESIDENT
> JACOBS ENGINEERING GROUP INC.
> PASADENA, CALIFORNIA

Integration

We are particularly interested in how companies of varying sizes effectively transfer a system of values from senior management to the hearts and minds of all key employees. This imperative seems to be growing even more complex these days due to the dramatic increase in the number of companies doing business in a variety of countries and cultures.

One example of particular interest is the architectural practice of Pringle Brandon Consulting in London. According to Jack Pringle, a director of the firm and 71st President of the Royal Institute of British Architects ('05–'07), "Only 35% of our staff is British. The great majority is from around the world." Naturally, we were interested in whether or not such an international group would have difficulty integrating a cohesive system of values throughout the office. "We live in a dynamic, multi-cultural city and our office reflects that. Perhaps that's why we pay extra attention to our group dynamic. The conversations are very rich," says Mr. Pringle.

So, why is it possible for Mr. Pringle to establish a focused team of professionals from such varied cultural backgrounds? It's time to go back to our friends in psychology, sociology, and organizational behavior. They point to research findings clearly demonstrating that when the tasks to be performed are complex, diverse work groups generate solutions that are more creative and innovative than groups comprised of culturally homogeneous participants. Further, we find research that compares the impact on job satisfaction of groups brought together by common demographics versus groups who share an overarching sense of shared values. Obviously, Mr. Pringle's staff doesn't share much of a common demographic and yet they function effectively as a team.

What they have is a unifying value system, which, as it turns out, along with a clear vision, are by far the most powerful tools available for creating a sense of purpose and focus within a group of professionals.

> Our competitors can talk just like us, but we have a 30-year head start in developing the values-based culture required to actually execute on a higher level.
>
> CHUCK L. GRECO, PRESIDENT AND CEO
> LINBECK GROUP, LP
> HOUSTON, TEXAS

The problem of integrating multicultural values seems even more challenging at Skanska, a 120-year-old, $17 billion, international construction firm. "We operate in 12 different home markets and because they are spread all over the world, each obviously operates in its own distinct culture," said Stuart Graham, President and CEO. "The challenge quickly becomes distinguishing between those values that apply only in a particular culture and those that we want to see adopted globally." Pursuant to their "decentralized but integrated" business model, Skanska has identified four values that all offices must actively pursue:

1. Zero environmental incidences
2. Zero loss-making projects—application of prudent risk management practices
3. Protecting the health and safety of employees
4. Zero ethical breaches

The executives at each business unit are annually evaluated in light of the above four values and the results impact incentive pay plans.

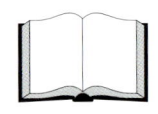

Value-Based Leadership: Rebuilding Employee Commitment, Performance, and Productivity
Thomas D. Kuczmarski
Prentice Hall, 1995

So, what causes an individual to accept or reject the values of any particular organization? In the past 30 years, a robust collection of theories has been generated in an attempt to explain the psychological underpinnings of human group behavior. Two of the more interesting ones include Group Identification Theory (Tolman, 1943) and the more recent Social Identity Theory (Tajfel & Turner, 1985). Both attempt to understand organizational socialization and human commitment or lack thereof in certain groups. While these approaches are still theoretical and often a touch academic, there are things practitioners can learn and apply in the real world.

First, social scientists explain that all humans have an insatiable need for self-acceptance and that one of the things people do to feel better about themselves is to affiliate with certain groups. An individual's decision to accept or reject the values of any given group seems to depend on several things:

1 Does the individual in question have a well-developed personal value system?
People with immature value systems will do almost anything for group acceptance while those who are more mature will be more discriminating.

2 How does the individual perceive the values of the prospective group?
Strangely enough, most individuals are not so concerned with how they perceive the group's values as much as they are concerned with how much social or professional status the group has in the minds of others.

3 Does the individual think the group's values match their own?
Once again, emotionally mature professionals are seriously concerned with a values match, while others simply want to feel better about themselves by associating with what they perceive as a high status group.

4 Does the individual believe that an association with this group will increase their personal or professional status?
This explains why we see so many people affiliating with sports teams, even though they don't play the sport, know any of the players, or even understand the intricacies of the game. They feel that wearing the paraphernalia creates an affiliation with an organization that embodies qualities the individual wishes he had.

Group Identification Theory and Social Identity Theory attempt to understand organizational socialization and human commitment or lack thereof in certain groups. While these approaches are still theoretical and often a touch academic, there are things practitioners can learn and apply in the real world.

Back to the original question: What specific steps can AEC organizations take to encourage a robust system of values among all of their workers?

Every organization we studied was serious about their core values and most were willing to invest energy in repeatedly discussing them. The great missing element, in most cases, was a specific program of evaluation and reward. If you're interested in actually impacting employee behavior,

you must specifically measure performance and publicly reward those who outperform the norm.

Otherwise, it's just talk.

Every organization we studied was serious about their core values and most were willing to invest energy in repeatedly discussing them. The great missing element, in most cases, was a specific program of evaluation and reward.

Evaluation and reward

At Gilbane, a 134-year-old construction and development company with 25 offices across the US, the topic of values is approached quite purposefully. They emphasize six categories:

1. Integrity
2. Tough-mindedness
3. Teamwork
4. Dedication to excellence
5. Loyalty
6. Discipline

A major element in their values integration process is Gilbane University, founded in 2000 by CEO Paul Choquette. The courses, 138 in all, are attended by every employee at every level of the company. Even though Gilbane U is an impressive effort, it is still perfectly natural for many employees to privately question the firm's actual commitment to stated values. It would be difficult, however, to ignore the fact that the course entitled "Living the Values" is taught by William J. Gilbane, President.

All of this is well and good, but Gilbane takes it a step further which is worthy of our attention: They evaluate how well their people live the values and they reward those who excel.

So, how do you tell who executes the values and who doesn't? Gilbane goes to the people that count—their customers. Each project must undergo an annual customer satisfaction survey specifically designed to rate staff on how well the customer sees them living the six core values. The results of the surveys are shared with all employees and they are a critical element in determining bonus payments.

It's important to have values and it's important to live by them, but a process of evaluation and reward adds teeth to the undertaking.

> **My performance rating is partially based on how many of my people attend Gilbane University and how well they perform in the courses.**
>
> WENDELL HOLMES, REGIONAL CEO
> GILBANE
> PROVIDENCE, RHODE ISLAND

Occasionally, we see a company willing to pass up financially lucrative opportunities that are at conflict with meaningful commitments to values:

Committing to values
Lance K. Josal, AIA
Senior Vice President
RTKL Associates
Chicago

Our firm was approached by a client to design a very large and prestigious project in China. The size of the project alone meant that the proposed projects' fee would be close to 5% of our firm's gross revenue for the entire previous year, not an insignificant amount for an architectural firm. After a few telephone conversations and a meeting with the client's representatives they finally came clean and told us that, "oh, by the way, the project design would need to have a themed design direction." Apparently, the client felt strongly that given the project location and the image they were trying to convey, a themed image would be the appropriate design avenue to follow to achieve their financial goals for the project. While this may have been a good business decision on their part, it triggered a corresponding business decision on our part. Most architects understand that theming is what you do when you don't have an idea and we felt we had plenty! So, their directive for the design esthetic posed a serious value question to our firm: were we more interested in a good fee or doing good design?

Included in my responsibilities directing this aspect of our firm's practice are the opposing goals of revenue generation and design quality, the former which is rather easy to assess (when in doubt, more is better) and the latter which is much more ambiguous (what is quality design?). I began by talking to my partners as well as associated project team members about the idea of getting involved in a project of this sort and, while they understood the fees in play, they were more than sanguine on the design approach. Left with a call to make, I felt this was not the type of project our firm ought to be designing. I felt that long after the fees had been bonused to our firm's employees, the design legacy would linger in perpetuity and hinder our growth and reputation more than even a significant fee could justify. I discussed the issue with my Chairman and President and they were supportive of my decision to decline the opportunity. I called the client and told them of my decision and wrote a long email with all the reasons why this didn't make sense for our firm. The client was respectful of our response, overly so actually, which

> caught me slightly off guard. Most important to me, though, were the comments from my partners and others copied on the email, all of whom were supportive, many of which stated they had "never before known our firm to do such a thing as turn down such a large project fee for philosophical design reasons!" Based on the reaction, the decision, it would seem, paid the firm immediate results and forever forward changed "the way we think of ourselves" as design professionals.
>
> While we passed on the previous opportunity, we value this client and they continue to avail our firm of other, non-themed opportunities around the world. They have come to respect our design capabilities for the inherent strengths we bring to the project team and we believe we can enhance our design reputation with these project opportunities as the client continues to offer us more, and most importantly, better opportunities to do the types of projects we both feel our firm is best suited to design.

In summary, what are today's market leaders doing about values?

Value Led Organizations
Eleanor Bloxham
Capstone, 2002

1 They believe
There exists a clear commitment to the idea that an overarching system of values can be a powerful force in organizing and focusing their efforts.

2 They talk
Values are openly and repeatedly discussed within the context of everyday project execution.

3 They train
Specific programs are created to formally explain their values, where they came from, and how values are expected to impact everyday performance.

4 They evaluate
Reliable methods for evaluating adherence are used and the results made public.

5 They reward
Employees are publicly recognized for exemplary performance and overtly paid.

6 They live
Hypocrisy is easy to spot. Value systems that aren't lived, everyday, by key people, will soon become marginalized.

In the end, companies that can capture a high level of job satisfaction and organizational commitment will outperform those that can't.

High-quality people want to work for a cause, not a company.

How will tomorrow's market leaders utilize values to be more competitive?

Tomorrow's market leaders will take what today's prominent firms have learned about values and simply turn up the volume. Instead of recognizing that a shared sense of values is an important part of the work experience, the next generation will address the fact that values and vision may, in fact, be the single most powerful organizing force available. We find that many of today's most interesting companies are experimenting with a variety of initiatives,

Tomorrow's market leaders will take what today's prominent firms have learned about values and simply turn up the volume.

all intended to more creatively apply values to their operations. We see advances being made on the following fronts:

Values discovery programs
If your company's system of values has been in place for many years, the issue becomes ownership. Tomorrow's leading firms will execute annual or semi-annual 'values discovery programs' that specifically engage all employees in thinking about which values will most effectively animate their work. The idea is to create a sense of ownership. When everyone shows up for work in the morning, they are there to execute their projects according to a value system they helped create.

Public status programs
The perceived status of your company not only impacts prospective clients, it materially impacts the quality of employees you attract. Everyone knows this. The point here is to not assume the public and professional image that your firm projects will occur naturally. Focus on it and do it on purpose. We'll discuss this more in the chapter on branding.

Values evaluation programs
If the two most powerful motivating and focusing tools you have are vision and values, why deal with them intuitively? And yet, that's exactly what most companies in the AEC industry do. The next generation of market leaders will regularly measure critical behavioral and attitudinal variables such as job satisfaction, organizational commitment, organizational citizenship, as well as typical performance measures.

Recognition programs
The idea is simple. Measure the behavior you want and publicly reward those who excel.

Project team value systems
In the AEC industry, people work in teams. Each team has its own organizational structure, power structure, vision, mission, and values. The characteristics of a well-functioning, values-based project team are easy to recognize:

- Clearly accepted mission
- Clearly understood expectations
- Ideas are synthesized, not 'sold' or 'dictated'
- Each member feels influential
- Each member feels accepted
- Commitment to process, not individuals or specific solutions

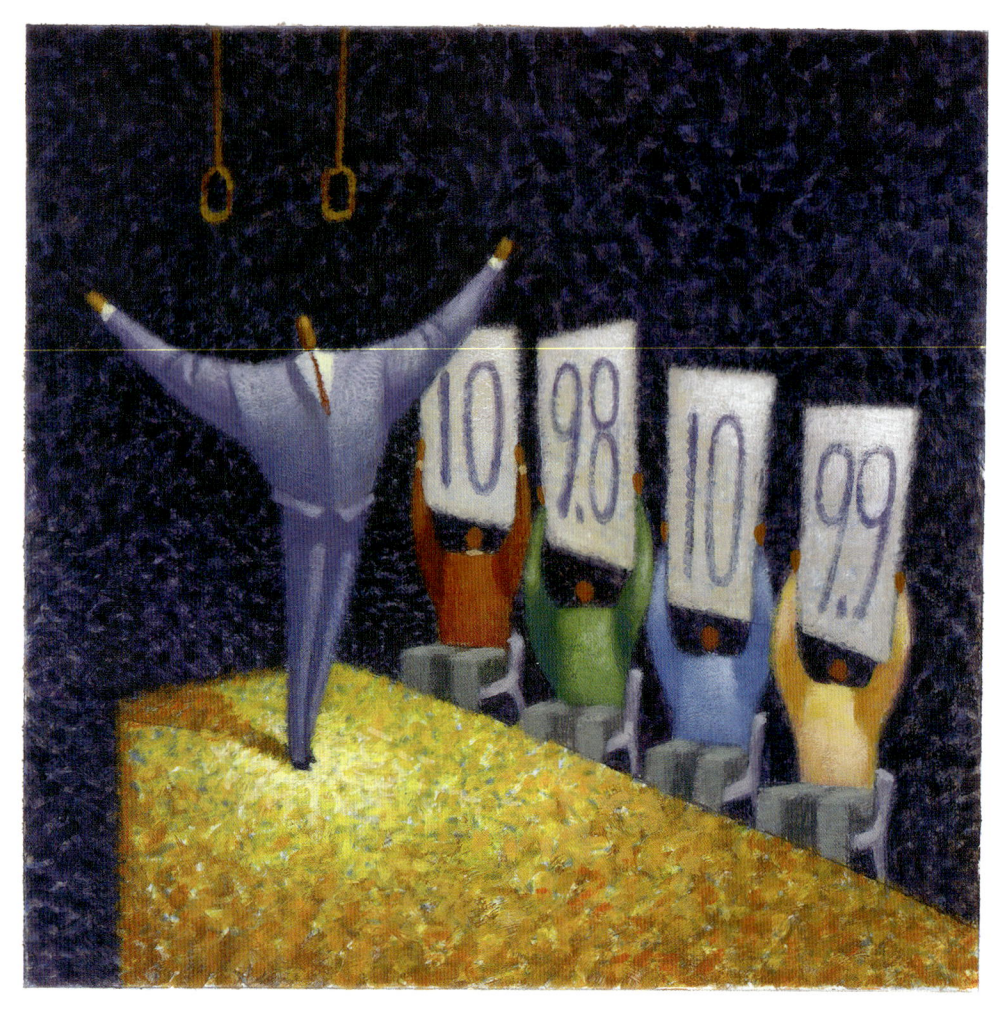

The next generation of market leaders will regularly measure critical behavioral and attitudinal variables such as job satisfaction, organizational commitment, organizational citizenship, as well as typical performance measures.

Tomorrow's market leaders will take the practices that are most effective at integrating values into their companies and apply them to each major project team in the office.

Integrated/interdisciplinary team value systems
Future designers and builders will concentrate on their capacity to function effectively in integrated teams of divergent professionals. Two years ago, we surveyed 98 large companies that own and operate substantial real estate portfolios. Our question: "When you create new facilities, what do you see as tomorrow's critical issues?" We were fully prepared to hear a panoply of complaints

about architects and general contractors. Instead, we heard this: "The processes used to create today's buildings take too long, cost too much, and, by the way, you architects, engineers, and builders don't integrate your expertise very well." That being the case, each of tomorrow's new integrated project teams won't function efficiently if it is not driven by vision and values.

> **M**ost of the professionals we interview for project teams talk about their values, but the truth is you don't really know who you're dealing with until a controversy occurs. That's when you learn what people are really made of and that's when you find your long-term partners.
>
> PETER HEALD, EXECUTIVE VICE PRESIDENT
> CGI DEVELOPMENT SERVICES
> LOS ANGELES, CALIFORNIA

Values champion programs

Since most values we cherish have a tendency to be somewhat ethereal in nature, one way to bring them alive is to assign specific "values champions." The idea is to honor individuals whose everyday performance embodies the group's most imperative values. We have seen groups designate "innovation champions," "integrity champions," "creativity champions," "process champions," etc. It's easy for this kind of thing to become trite, but properly and earnestly executed, it can be a powerful technique for keeping values at the top of the agenda.

Value integrated branding events

Most firms participate in a series of professional and community events designed to enhance the company's public image. The power here is not just the imperative of taking your message to the business community. You are talking to your own employees at the same time, and it can have valuable benefits.

The clearest example of this phenomenon occurred about 25 years ago when the United States Postal Service spent millions on a national TV ad campaign. Why would they do that? They are a government monopoly and at the time had absolutely no competitors. This was a truly unusual occurrence. Every other public message, regardless of the content, when promulgated by a private sector company, has the intention, one way or another, of impacting revenue. Since revenue and profit are both irrelevant to a government agency, what were they up to?

As it turns out, the Postal Service had data saying that their six million employees were, as a whole, not proud of their jobs. This was impacting productivity, turnover rates, retention rates, and the quality of person that the Service could attract to work for them. People will not adhere to the value system of a group they do not admire.

Instead of talking directly to their employees, management decided that step one was to elevate the image of their company in the minds of all Americans.

It worked. The ads ran for two years and successfully reversed the organizational maladies caused by being perceived as a low-status place to work.

The rewards for being perceived as a high-status company are substantial, even if most measurements of that condition are actually irrelevant. As a younger person, I got a chance to look inside several prestigious organizations only to find that they were, at best, completely common.

Back to my point. Even if you know that many status and prestige measures are wrong-headed, play the game anyway. It's part of communicating your values to not only the marketplace, but to your own staff as well.

We will discuss this further in later chapters, but it's impossible to overemphasize the power of branding your values.

References

Tajfel, H. and Turner, J.C. (1985) Psychology of Intergroup Relations. In S. Worchel & W. G. Austin (Eds.), *The Social Identity Theory of Intergroup Behavior*. Chicago: Nelson-Hall.

Tolman, E.C. (1943) Identification and the post-war world. *Journal of Abnormal and Social Psychology*, **38**, 141–148.

Competitive Focus

This chapter is not about how to fabricate a competitive strategy (there are libraries about that already). This chapter is about defining, creating, and maintaining competitive focus.

Chapter 3

COMPETITIVE FOCUS

What's so critical about competitive focus?

What's wrong with how we currently plan?

How will the next generation of market leaders use competitive strategic planning to gain market prominence?

In the early 1960s, a bright shiny new management discipline arrived entitled "competitive strategic planning." What a fabulous idea. Strategy would no longer be the intuitive, commonsensical backbone of every successful venture. It would hereafter become a formalized conspiracy of experts, complete with an impressively stylish new lexicon, computer modeling, pricy consultants, and reserved parking for its most fashionable practitioners. Their charter was clear. First, pull together a collection of MBAs, PhDs, financial experts, marketing gurus, and assorted rodeo clowns. Next, have them analyze all worldwide industry variables, both real and imagined, then plot an unassailable course to corporate success. What fun.

The work done by these staffs of strategic planners was designed to lead to a new and startling level of corporate rectitude. The right products and services would be introduced to the right markets at the right time. The results of these strategies would create a competitive juggernaut, the envy of all the corporate kingdom.

It didn't work.

By the mid-1980s, study after study revealed that companies employing formal strategic planning staffs did not outperform companies that had no such function. In 1983, Jack Welch, Chairman at the time of General Electric, dismantled his company's highly-regarded 200-person strategic planning group, stating that they were out of touch with the realities of day-to-day competitive forces.

In hindsight, the most compelling reason for the failure of formal competitive strategic planning seems simplistically apparent: Strategy has no business being developed by a bunch of eggheads secluded on the executive floor of the corporate headquarters building.

It must thrive in the heart of the company, and therefore, should be born and nurtured there.

Our investigation of strategy and how it impacts competitiveness has led to several powerful observations. First, every company in the AEC industry genuinely believes they are fully engaged in creating and executing a competitive strategy. Typically, what they actually have are pieces of unaligned initiatives strewn about the office; some elements are new and exciting, others are old and moldy.

Competitive Strategy Dynamics
Kim Warner
John Wiley & Sons, 2002

> This chapter is not about how to fabricate a competitive strategy (there are libraries about that already). This chapter is about defining, creating and maintaining competitive focus.

The game is changing

Our work has led to one observation that immediately jumps off the page. The competitive strategies that propelled last generation's market leaders simply don't have the horsepower to do the same for the next generation. The game has become measurably more complex.

Historically, market leaders concerned themselves with four things: differentiation in services, differentiation in credentials, personal networking, and persuasive competitive selling.

I was 26 when I told my first boss that I wanted to get involved in new business development. He suggested that I attend at least one socially impressive church, join several civic clubs, learn to play golf, and marry into a well-placed family. He wasn't kidding. I then asked if we had any firm brochures I could hand out. He said that we did, but they cost too much to give away to anyone not planning a major construction project. The idea at the time was that industry leaders needed to be competent professionals and socially connected. While it never hurts to know people, the next generation is going to have to approach things more intelligently.

> **E**very year we rethink our competitive focus statement. When an opportunity arrives with circumstances outside our focus, we require board approval before pursuit. Our focus statement clearly delineates the opportunities we intend to pursue, along with specific client and size requirements.
>
> Over the years, we have become committed to an integrated partnering approach that demands that we pursue relationships, not individual projects.
>
> SIR MICHAEL LATHAM, DEPUTY CHAIRMAN
> WILLMOTT DIXON CONSTRUCTION
> LONDON, ENGLAND

The Rise and Fall of Strategic Planning
Henry Mintzberg
Free Press, 1993

Here's the problem. Generic service differentiation is very difficult to achieve, even more difficult to maintain, and often complicated to convincingly sell. Credential differentiation is the same story. Most builders, engineers, and architects can hardly wait to get in front of a new prospective client, get them in the conference room, turn down the lights, fire up the visuals, and describe with senile rapture the full breath of their impressive credentials. While this is big fun for the presenter, it's of limited value in getting to the finish line.

Ask any repetitive developer of new buildings and they will tell you that the credentials of the top five competitors are virtually indistinguishable.

What's so critical about competitive focus?

No company has enough resources to pursue every course that seems attractive. Even the largest companies must focus their resources to create the greatest impact. This means effective planning. It also means responding to genuine opportunity when appropriate. The fun starts when the latest fad comes knocking on your door dressed exactly like an exciting opportunity. Or is it an actual opportunity dressed like a fad? If you pursue it and it's a fad, substantial resources will be wasted. If you don't and it turns out to be a long-term trend, then you can find yourself late to the game and lose serious ground to some lesser firms who happened to guess right. There's a lot at stake when a company of any size sits down to decide where and how to compete.

No company has enough resources to pursue every course that seems attractive. Even the largest companies must focus their resources to create the greatest impact. This means effective planning.

Competitive Planning at BDP
Peter Drummond, Chief Executive Officer
Building Design Partnership
London

Design businesses prepare plans all the time. It's what we do—making plans for cities, buildings, spaces, interiors, structures, building services. We're also pretty good at self-analysis, in a self-conscious, creative and intuitive kind of way. And so making plans for our own businesses should therefore be a straightforward and commonplace process. But I imagine that the truth is that our focus on client service, the next project, design quality, resource management and getting things built understandably takes most of our creative energies, and our businesses evolve organically and responsively.

At BDP, we have tried to place our vision and plan for the next five years in the hearts and minds of the business and all of our people. The Plan, as it is affectionately known, is the basis of financial business planning and decision taking, a clear statement about our aims and values, and a set of actions (known as strategic imperatives) for all the main operational parts of the firm.

The Plan was prepared over a six-month period, involving debate and discussion with 150 members of BDP, facilitated by a firm of strategic management consultants, covering every region, profession, business sector, and enabler. Once completed, it was communicated to all parts of the firm by the Chairman and Chief Executive, is available on the intranet, and all members and new starters in the firm are given a published booklet.

The Plan starts with a vision of BDP in five years' time:

- "an irresistible firm, consistently designing and delivering outstanding buildings and spaces
- a £110 million business, operating profitably, efficiently and with great spirit
- world leaders in interdisciplinary design of buildings and spaces which grace their environment, create real value for our clients, and respond to the needs of those who use them every day
- a place where the principles of teamwork, collaboration, innovation, and creativity thrive."

A core part of the Plan is an honest assessment of strengths and weaknesses, along with an analysis of the breadth of opportunities areas to address. This included a reworking of BDP's core positioning logic—**design integrity**—a core value that informs every aspect of who we are, what we do, and how we do it. **Design integrity** symbolizes the clarity of BDP's purpose, that of the highly-skilled conception, creation of plans for great buildings and spaces; and the benefits that flow to our clients—through delivery of a product that embodies soundness, completeness, and unity.

Financial objectives were defined for all parts of the business, and these are used to drive the annual budget-making process. These, along with the strategic imperatives, are monitored and assessed on an annual basis by the BDP Executive Board and progress reported to the entire firm via a twice-yearly e-newsletter. Two years in—so far so good!

As with all good plans, the BDP Plan will change well before the five years are up. This is viewed as an opportunity rather than a chore—an opportunity to debate, consult, envision, and set the path for the next five years.

So what can we learn from firms in the AEC industry that have successfully navigated the uncertainty? And what can we learn from those that haven't?

Ten Rules for Strategic Innovators
Vijay Govindarajan and Chris Trimble
Harvard Business School Press, 2005

So what can we learn from firms in the AEC industry that have successfully navigated the uncertainty? And what can we learn from those that haven't?

> Before a small but growing practice like ours can distinguish an opportunity from a distraction, it must go through a process of generating a clear, executable vision.
>
> STEPHANIE FORSYTHE, DESIGNER
> MOLO
> VANCOUVER, CANADA

What's wrong with how we currently plan?

Later in this chapter we will seriously challenge current strategic planning conventions, but for now, consensus clearly falls in favor of rigorous competitive strategic planning. That being the case, our investigation reveals that planning failures and shortcomings occur for reasonably predictable reasons.

1 Based on common data

Every strategy is based on human opinions which are supposedly based on objective information. It's not unusual to find many of the firms in any market area basing their plans on very similar data. Similar data makes for similar assumptions. Similar assumptions will make your strategy look and feel just like everyone else's. The first and perhaps most critical step in developing a competitive advantage is to acquire data from a deeper and richer source than your competitors. We have observed a couple of interesting sources for compelling market knowledge: (1) your own employees, (2) your best clients (involve them in planning your strategy), (3) targeted prospective clients (they won't all talk to you, but a few good ones can be extremely valuable).

2 Takes too long to develop

We found several firms that had spent years developing the intricacies of their strategic planning process. The result was a formality that can't suitably respond to current market dynamics. These planning processes often serve the business requirements well, but aren't a good environment to nurture new, creative competitive strategies.

3 Process not iterative
A perfectly frightening thing to hear from any company is that "our competitive strategic plan is finally finished." A more interesting approach is to see all strategy decisions as experiments, trials, or tests. We will discuss this idea later in this chapter.

4 Uninspiring
Humans are emotional creatures. Any new strategy, program, or initiative that doesn't inspire simply won't get the kind of energy required for high achievement.

Strategies based solely on a rational analysis of market data will probably fail to ignite the level of emotional attention necessary to excel in a highly competitive arena.

Eric R. Kuhne, with Civic Arts in London, a man of great passion about his work, inserts ideas in his firm's competitive strategies specifically meant to evoke an emotional response. "We pursue projects and clients that provide us with a new pair of eyes through which to see the world." This phrase is obviously not measurable, but it was never intended to be.

5 Not integrated into the group consciousness
Several of the companies we studied were perfectly adept at creating serviceable competitive strategies and formally announcing those strategies to their people. Three months later, when we asked the key employees about the company's competitive strategy, most of them couldn't remember it.

6 Not translated into measurable missions and goals
Success is about execution. It is not unusual to find companies with perceptive, interesting strategies that don't impact behavior. Generally this occurs because the strategies were never translated into specifically measurable goals. I hate to be repetitive, but you get the performance you measure and the performance you pay for. If you don't measure it and specifically pay for it, it's just talk.

Execution: The Discipline of Getting Things Done
Larry Bossidy and Ram Charan
Crown Business, 2002

7 Don't have the right people
Testing new market opportunities with the wrong people wastes time and valuable assets. This isn't news and so commonsensical as to not be worthy of much discussion. So why do we see it so often?

> When we're considering the pursuit of a new opportunity, naturally we filter it through our mission statement. But there is a more important consideration: Does anyone in our group want to own it? If the answer is no, we don't proceed no matter how well it matches our existing strategy.
>
> LESLIE BUTTERFIELD, CEO
> MCLACHLAN LISTER PTY LTD
> SYDNEY, AUSTRALIA

8 Culture not suitably agile

Every AEC firm on the planet operates under a set of values and an operable group culture. Some of these systems were developed intuitively while others were consciously fabricated. Agility, which we will discuss more fully in later chapters, is not a function of company size but culture. As the future becomes more and more difficult to forecast, agile cultures become more and more valuable.

Most of the cultures we studied actually serve to retard agility rather than empower it.

Ben Skerman, Queensland's 2006 Young Professional Engineer of the Year, spends his waking hours pursuing brand extension projects. His employer, WorleyParsons, is fully aware that the new work Mr. Skerman is chasing can't be accomplished without new and innovative organizational responses. "Agility is a value we cherish and it is purposefully designed into our organizational structure," says Mr. Skerman.

The Power of Strategy Innovators
Robert E. Johnson and
J. Douglas Bate
AMACOM, 2003

9 Inadequate focus

We witnessed one medium-sized, well-respected architectural firm that had identified 16 new market initiatives to pursue … too many targets, not enough ammunition.

> Skanska is known as a global construction company, but actually we are a multi-local company that focuses entirely on each of our home markets, outside of which we don't wander.
>
> STUART GRAHAM, PRESIDENT & CEO
> SKANSKA
> STOCKHOLM, SWEDEN

How will the next generation of market leaders use competitive strategic planning to gain market prominence?

The last generation typically used its strategic planning process to predict the future as accurately as possible and plot a course to sustain specific markets and attack others.

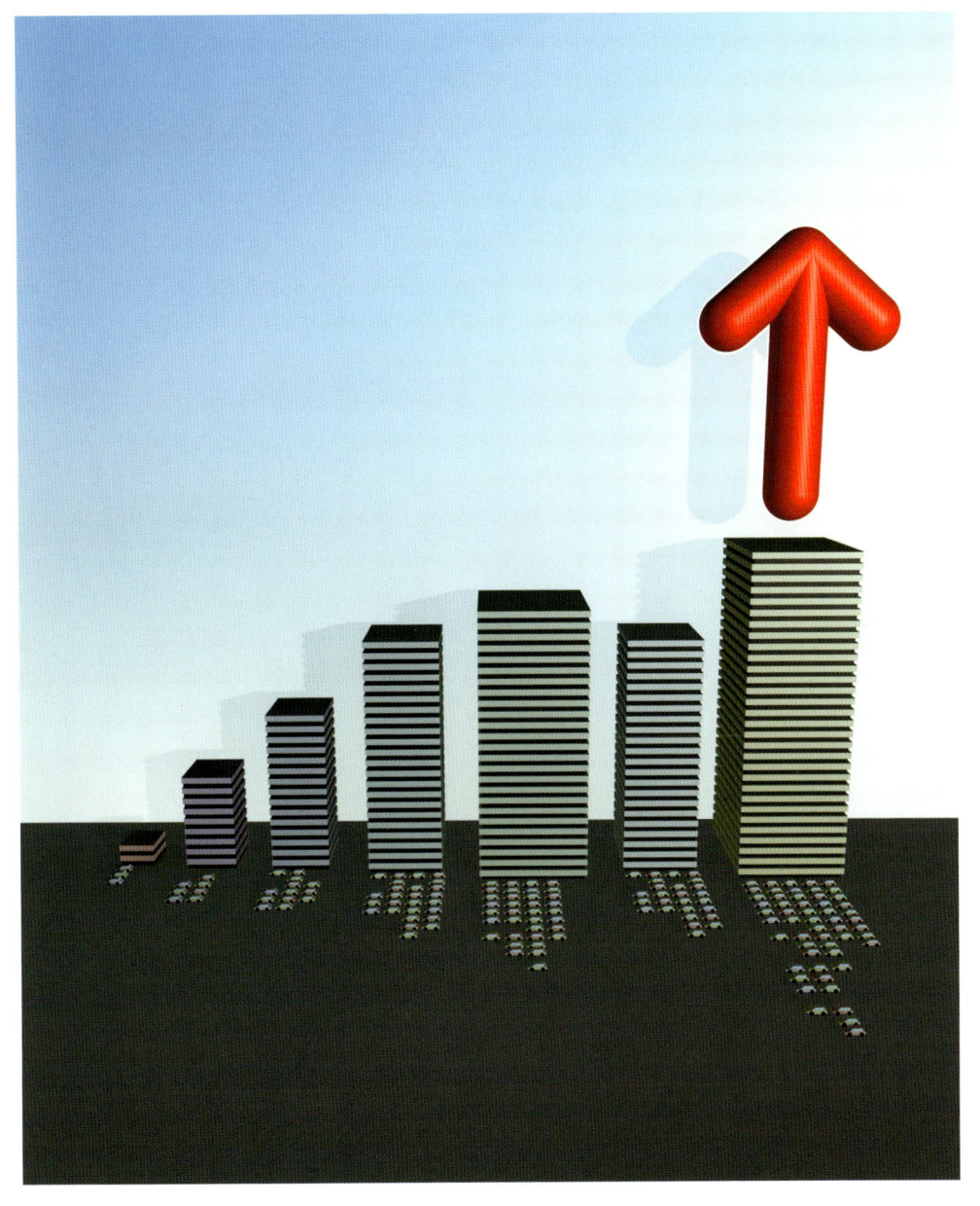

How will the next generation of market leaders use competitive strategic planning to gain market prominence?

Competitive Focus

Our work has identified two problems with this traditional approach that the next generation of market leaders must work to rectify.

First, we have witnessed the process of strategic business planning degrade from an energetic, creative undertaking into a rigid affair that serves to deal primarily with financial allocation requirements, budgets, schedules, personnel assignments, etc. While valid business issues can be served by this kind of annual business planning, we believe that the process used to envision new competitive strategies needs to take place in a more creative, inclusive, experimental, and iterative environment.

Second, the old business planning model requires that the participants accurately forecast relevant market variables such as: "What will our customers be demanding and how will our competitors be responding?" If global competitive forces continue to vibrate with increased uncertainty, traditional planning will become less and less relevant.

Let's recap: We're concerned with traditional planning because it has become too formulaic and it requires forecasts that are becoming less and less precise.

So, does anyone have a better idea? Our findings suggest revisions that address both problem areas:

The old business planning model requires that the participants accurately forecast relevant market variables such as: "What will our customers be demanding and how will our competitors be responding?" If global competitive forces continue to vibrate with increased uncertainty, traditional planning will become less and less relevant.

1 Separate the competitive strategy process from traditional business planning.

Since formalized business planning wasn't producing competitive advantages 30 years ago when the world was substantially less chaotic than it is today, how can we expect it to yield acceptable results now? We can't. When the time is appropriate, you can fold the results of your competitive strategy planning back into your annual business plan.

Designing a stand-alone competitive strategy curriculum will allow you to inject new and more perceptive sources of data:

(a) All key employees

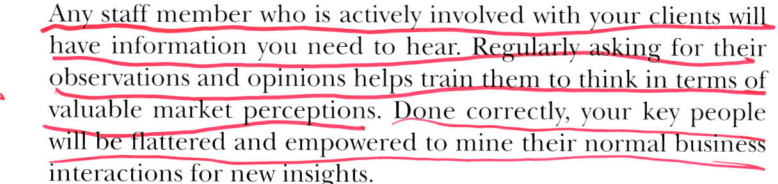

Any staff member who is actively involved with your clients will have information you need to hear. Regularly asking for their observations and opinions helps train them to think in terms of valuable market perceptions. Done correctly, your key people will be flattered and empowered to mine their normal business interactions for new insights.

(b) Selected existing clients

As long as you don't demand too much of their time, some of your most valuable clients will agree to participate in your competitive planning program. They will be impressed with your interest in being more valuable and may supply far more perceptive intelligence than your competitors will have access to.

(c) Selected prospective clients

We even witnessed several firms who invited high-value clients, with whom they don't currently work, to briefly participate in a "future visioning" process. Often these people genuinely enjoy the interaction with other industry leaders and will occasionally share unusually powerful information.

2 Don't ask your strategy process to produce a specific course to follow.

Your planning process simply can't optimize your competitive position in an environment that appears materially unsure. The next generation's most effective companies will use their competitive development process to produce, not a list of specific answers, but a list of creative ideas worthy of testing. The best ideas will be ranked and company assets assigned accordingly.

All new competitive ideas will be seen as trial initiatives to be constantly nurtured and appraised.

> **P**lanning creates plans, not strategy.
>
> ———
>
> HENRY MINTZBERG, ASSOCIATE PROFESSOR
> MCGILL UNIVERSITY
> MONTREAL, CANADA

20/20 Foresight:
Crafting Strategy in an
Uncertain World
Hugh Courtney
Harvard Business
School Press, 2001

Management at HKS, the largest architectural firm in Texas, annually revise their three-year competitive strategic plan. The interesting thing about their approach is that each of their designed strategies is seen as a "program of transition," says President and CEO Ralph Hawkins. "Over time, everything internal and external to our company is changing in form, nature, and content. All of our clients are evolving as are our people. This puts new and interesting demands on the assets and support systems of our company. Every functional element we have is in transition. We, therefore, see our competitive planning process as a series of steps that must be constantly monitored and evaluated."

Category Ownership
What does it mean to own a category?

Chapter 4

CATEGORY OWNERSHIP

What does it mean to own a category?

How did today's category owners achieve their positions?

What role does customer intimacy play?

How will the next generation develop and exploit new categories?

Shortly after the end of World War II, the world seemed, somehow, more orderly, and Americans finally had the opportunity to turn their collective attention to matters of greater impact than world peace ... college football. In the American Southwest, Texas A&M University had a running back who they felt was worthy of winning the highest individual award in the sport, the Heisman Trophy. The prize is given at the end of the season and was originally chartered to recognize the best college football player in the nation ... a ludicrous idea on the face of it. Since there are approximately 15,000 Division 1A college football players, it is obviously impossible to rationally select only one for this impressive recognition. But in a media-driven culture, it's important to invent heroes, even if the process makes no sense.

That last comment was a personal rant. I apologize. Back to my story.

The Heisman typically goes to a player with impressive statistics—most yards gained, most touchdowns made, most passes completed, etc. The university's problem was that their player, John David Crow, was an exceptional athlete, but not the best by any traditional measure. He hadn't gained the most yards or scored the most touchdowns, and yet they still wanted to promote him as the best runner in the nation. So, they did the creative thing. They invented a new category ... one that their player could own—most yards gained after initial contact. No one had ever considered such a statistic, so when the university announced it, they also proudly proclaimed their player to be the best in the nation at running after being hit.

I'm certain you've guessed by now the purpose of this story. John David Crow did, in fact, win the Heisman Trophy, an accomplishment that was clearly not possible had Texas A&M promoted him according to traditionally accepted statistics. They invented a new category and took ownership.

Typical categories

The AEC industry, worldwide, functions within a complex tangle of project categories.

High-rise	Mixed use
Healthcare	Master planning
Retail malls	Urban & regional planning
Retail strip centers	Airports
Hospitality	Urban transportation
Sports venues	Corporate offices
Warehouse	Manufacturing
Education K-12	University
Museums	Performance venues
Research labs	Government
Convention venues	Courthouses

Momentum: How Companies Become Unstoppable Market Forces
Ron Ricci & John Volkman
Harvard Business Press, 2002

What does it mean to own a category?

We all know that within each of these project categories, or market segments if you prefer, there are firms that have found a way to achieve a level of prominence. They are easy to identify and we all know who they are. Ask any major buyer or influencer who the top five firms are in any particular segment and they will instantly recite the names. This is obviously enviable territory for any company to inhabit. It means that when new work is being envisioned, the prominent firms will almost always be invited for discussions. It makes sense, then, that these companies have a substantial investment in keeping the market segment descriptions just the way they are. The system is working wonderfully for them.

But what if your firm is currently not a category leader, not one of the names that instantly comes to mind? Taking on a collection of entrenched, prominent firms can make for a long day at the office. It means you're choosing, of your own free will, to play another firm's game, in another firm's stadium, by another firm's rules, in front of another firm's fans. It also means that you're going to do more than your share of losing for many years while you learn to beat very good companies at their own game. While younger, smaller, more aggressive firms do, in fact, from time to time capture the high ground from older, more credentialed competitors, the process is usually laborious and fraught with failure. It is critical that we come up with a better, faster way.

The most powerful path to prominence within existing AEC project categories is to stretch some, shrink some, reconfigure some, completely invent new ones and, shortly thereafter, plant your company flag squarely in the center of the newly devised market segment and declare ownership.

Common knowledge

Those of us who are aggressive consumers of business literature know all too well that most business books waste untold pages explaining the obvious while illuminating with senile delight things everyone already knows. I actually read recently: "A new university study has found that people who consume fewer calories gain less weight." While I'm certain that the study was executed to the highest academic standards and the results were statistically significant, the findings are common sense to everyone on the planet older than six. Since it is important for us not to bore you with conclusions of which you are already aware, here's an overview of what is considered common knowledge about the process of reconfiguring market segments for fun and profit.

1. Market segments exist because they make sense in the minds of most buyers. You must deal with them.
2. Competing in established project categories with established firms is a bad idea.
3. Small, more energetic companies are typically better at ferreting out and defining new categories and new approaches.
4. Once new categories are illuminated, larger firms are typically better at scaling up the new idea and commercializing it on a large scale.
5. Attempting to invent entire new industries is an incredibly difficult task. These are often referred to as 'blue sky' or 'green field' or 'blue ocean' strategies. Why the necessity for color descriptors, I don't know. It happens, but not very often.

Competitive Dominance: Beyond Strategic Advantage
Victor Tang and Roy Bauer
John Wiley & Sons, 1995

The more effective way to go about this adventure is to work with existing market segments and modify them to your advantage.

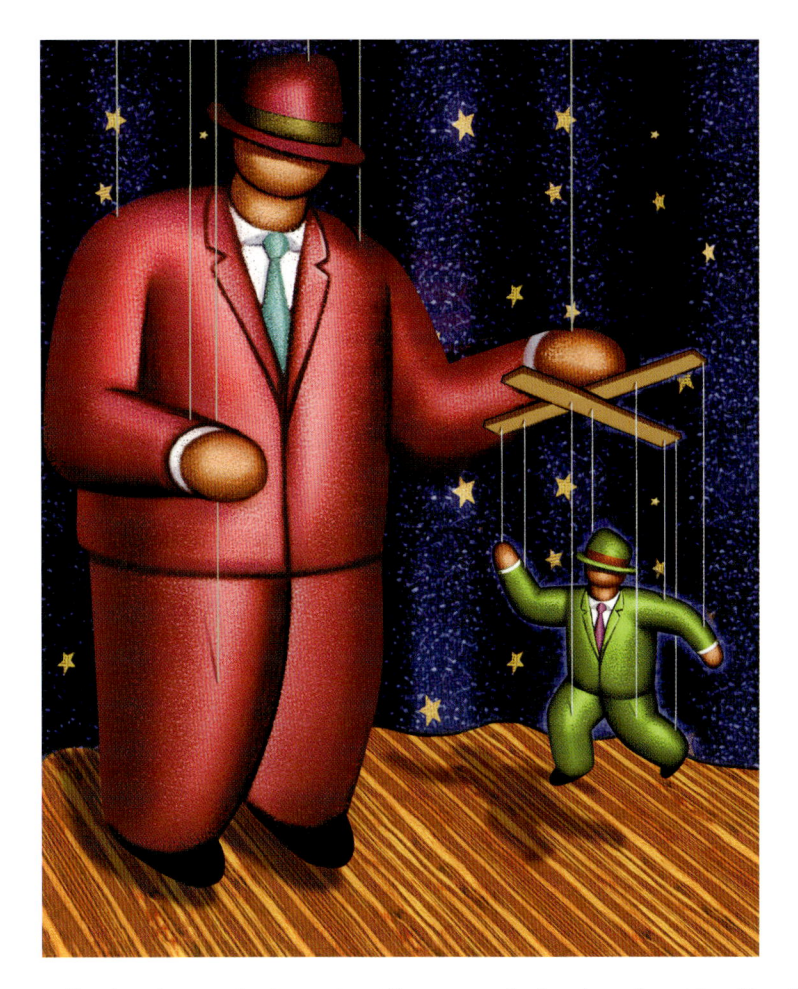

Taking on a collection of entrenched, prominent firms can make for a long day at the office. It means you're choosing, of your own free will, to play another firm's game, in another firm's stadium, by another firm's rules, in front of another firm's fans. It also means that you're going to do more than your share of losing for many years while you learn to beat very good companies at their own game.

How did today's category owners achieve their positions?

There are obviously infinite pathways to the goal of **market segment redefinition**. Some companies just happen to be in the right place at the right time when something they're good at gets unexpectedly popular. Others are dragged into new lucrative business opportunities and forced to make money, against their will, by persistent clients. But the events we study are the ones where perceptive competitors execute a consciously designed program to capture and own new market segments.

> **B**y making the change in our office, 15 years ago, to category specific teams, we have been able to grow our firm with targeted marketing pieces and team leaders that focus solely on their market type.
>
> JOHN KIRKSEY, PRESIDENT
> KIRKSEY ARCHITECTURE
> HOUSTON, TEXAS

Here's a brief description of strategies that we've witnessed producing repeated success:

1. Apply an existing service in a new sphere
2. Define process as a category
3. Identify an underserved market segment
4. Create a joint venture between an innovator and scaler
5. Research and publish
6. Specialize more deeply

1 Apply an existing service in a new sphere

It's not something to count on, but occasionally providence smiles on us. The question quickly becomes, what do we do now? How do we capitalize on newly presented opportunities? Our experience is that most AEC firms don't energetically and purposefully exploit opportunities that the marketplace occasionally provides. When presented with a new, out-of-the-ordinary project, most simply do it and go on to the next one.

Others, however, make a habit of considering any assignment requiring a professional stretch as perhaps a window on a future lucrative market.

The single most common event that propels a company toward prominence in a new market category is being asked by a client to apply an existing skill set in a different way.

It's happened several times at Drivers Jonas, a seven-hundred person commercial property consulting firm headquartered in London. Among other things, they are highly regarded for their project management expertise and were asked about 12 years ago to manage the development of two sports stadia in the north of England. These new projects represented an opportunity for Drivers Jonas to apply their existing project management skill sets in a new domain, and the company made the conscious

The single most common event that propels a company toward prominence in a new market category is being asked by a client to apply an existing skill set in a different way.

decision to use this opportunity to define a new market category and declare ownership.

They created a team of specialists, provided them with the new moniker, "DJ Sport" and began developing expertise in the financing, planning and delivery of sports facilities. The company successfully maintains the attention of major sport facility buyers by creating a steady stream of publications, industry surveys, and educational seminars. Today, DJ Sport provides a substantial source of company revenue.

Category Ownership

> The process of specializing in market categories worked well for us in sports, but it is equally effective in education and healthcare, where businesses happen to have a lot of property, but are not in the property business.
>
> NICK SHEPHERD, MANAGING PARTNER
> DRIVERS JONAS
> LONDON, ENGLAND

A similar set of events occurred at Walter P Moore, a large, integrated national engineering and consulting firm with ten offices throughout the US and headquarters in Houston. Founded in 1931, the firm has built a reputation for being able to handle difficult technical problems. It was therefore selected, in the mid-1960s, to do the structural engineering for the Astrodome, the world's first domed sports arena. It was designed for football, baseball, rodeos, concerts, destruction derbies and any other activities humans could imagine requiring grand volumes of indoor air-conditioned space.

As a side note, the Astrodome also provided a new brand and a substantial source of revenue for Monsanto Chemical, which was asked to develop plastic grass when the real stuff promptly died upon installation. The turf specialists had hoped that the roof's skylights would provide enough sunlight for the real grass to flourish, but alas, it didn't. The skylights were summarily painted, the botanical grass replaced by plastic, and everyone lived happily for the next 40 years until sporting experts decided that Astroturf was the groundcover equivalent of disco music and polyester leisure suits.

In the 1990s, however, the trend in sports venues became nostalgic and moved away from multipurpose indoor facilities to freestanding, single-user buildings. That means the baseball team and football team now want separate fresh-air stadiums. Well, not exactly. They actually want outdoor facilities when the weather's nice and indoor arenas when the weather's inclement. And, by the way, they want to be able to make the change in ten minutes.

With Walter P Moore's structural credentials, it was selected to design a baseball stadium with a moveable roof, Minute Maid Park in Houston. Once again, instead of considering this project just as an interesting, new assignment, the firm set in motion the process of defining retractable roofs as a new market category, one it could dominate. Over time, it established a specialized R&D group to explore new materials and capture the best thinking, worldwide, on the subject.

The world's first—and largest—football arena with a retractable roof, Reliant Stadium, was Walter P Moore's next big assignment and the engineers were well on their way to being considered the category leader, an event not animated by happenstance but purposeful planning.

> **O**ur sports stadia clients are looking for exhilarating, functional new facilities, but at the same time they must find a way to mitigate the risk of cutting-edge structural technologies. Our level of specialization allows owners to capture the excitement of retractable roofs while minimizing the risk.
>
> RAYMOND MESSER, P.E., PRESIDENT AND CHAIRMAN
> WALTER P MOORE
> HOUSTON, TEXAS

Hidden in Plain Sight: How to Find and Execute Your Company's Next Big Growth Strategy
Erich Joachimsthaler
Harvard Business School Press, 2007

2 Define process as a category

In the AEC world, when we think of market categories we typically think in terms of building types: healthcare, hospitality, retail, high-rise, education, etc., but it doesn't have to be that way. Linbeck, a well-regarded Houston-based construction firm with five offices around the US, decided some years ago that they would not pursue differentiation by specializing in specific building types, but distinguish themselves with any type of complex building by promulgating a more efficient project delivery system. Their approach came about in an interesting way.

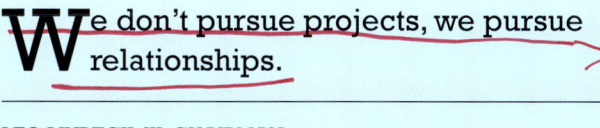

> **W**e don't pursue projects, we pursue relationships.
>
> LEO LINBECK, III, CHAIRMAN
> LINBECK
> HOUSTON, TEXAS

In the late 1960s, Linbeck successfully built a 35-story office building in downtown Houston for a prominent local real estate investment firm. Five years later, the developer called the President of the firm, Leo Linbeck, Jr., with an interesting request: "We want another building, just like the first one you built for us, at the same price." Linbeck promptly fished out all original drawings and specs and repriced a new building. Obviously, all new costs now had five years worth of escalation living in them, making the goal of creating another building at the same price out of the question.

Mr. Linbeck then proposed a creative idea, which later became the basis for defining a new market category: He explained to his good client that he could build another building that looked and functioned just like the original, at the same price, if they agreed to let Linbeck control the process.

> **Process doesn't sell, results do.**
>
> ―――――――――――――――――――――
>
> MELVIN HILDEBRANDT, FORMER PRESIDENT
> LINBECK
> HOUSTON, TEXAS

The developer agreed. The new building came in on time, under budget, and a new market category was born—driven by process not building type. Linbeck named this new proprietary project delivery system TeamBuild® and it continues to serve as the backbone of their 100% negotiated planning and construction practice.

3 Identify an underserved market segment

Thirty years ago, the commercial real estate consulting industry was dominated by professionals whose job was to represent sellers or landlords or building developers, the companies that own or create inventory. The users of these buildings, some buyers, some tenants, often found themselves negotiating on their own behalf without a full understanding of all real estate options available to them. A couple of young real estate entrepreneurs in Houston recognized what appeared to them as an underserved category: tenant or user representation.

Gerry Trione and Charles Gordon formed their new firm, and over a period of 20 years, methodically executed a three-phased plan:

1. Identify an underserved market segment and take ownership
2. Expand and reconfigure that segment based on an unusually intense level of customer intimacy
3. Inflate their market reach to the national level

Their new venture, Trione & Gordon, began with the traditional, if underutilized, service of tenant representation, but had their efforts ended there, a new and powerful new market category would never have evolved. "Led by our largest clients, we quickly determined that while negotiating an optimal lease is important, the real issue is how to efficiently house their employees," says Trione. "Those sets of services then led us to the financial world

Thirty years ago, the commercial real estate consulting industry was dominated by professionals whose job was to represent sellers or landlords or building developers, the companies that own or create inventory. The users of these buildings, some buyers, some tenants, often found themselves negotiating on their own behalf without a full understanding of all real estate options available to them. A couple of young real estate entrepreneurs in Houston recognized what appeared to them as an underserved category: tenant or user representation.

that underpins all real estate decisions. We now spend most of our energies designing new financing alternatives." Each step along the way, Trione & Gordon, led by a deeper understanding of their clients' real estate needs, morphed its services to become more valuable and more relevant.

The lesson here: Identify an underserved market segment, declare ownership, and let your best clients lead you to a redefined, more valuable market category.

The Innovation Equation: Building Creativity & Risk Taking in Your Organization
Jaqueline Byrd and Paul Brown
John Wiley & Sons, 2002

So for Trione & Gordon, the next step was to take the market prominence that they acquired in the fourth-largest city in the US and escalate it to a national or international scale, a difficult assignment indeed for any company that provides an intensely customized service.

4 Joint ventures between innovators and scalers

As mentioned in the 'common knowledge' section of this chapter, most innovation occurs in small, high-energy, risk-accepting firms. But seldom do these creative pockets of professionals have the market muscle to amplify their successes to national or international scope. For that, a partnership with a larger group sometimes makes sense for both.

Back to our story about Trione & Gordon. At this point they have identified an underserved market segment, aggressively reshaped it to their advantage, and in the process created a name for themselves as the leading tenant representation firm in Houston. Even with this level of success, the task of translating their local prominence to a national scope seemed, indeed, Sisyphean.

Into our saga now arrives Will Penland, Senior Managing Director of the largest commercial real estate services company in the US, CB Richard Ellis, Inc. (CBRE), with offices in every major American city, didn't acquire its lofty position by trailing market expectations. They expect national prominence in each of their three major organizational categories: (1) real estate development, (2) property management, and (3) real estate services.

> **When you talk about owning a category, you talk about making and keeping star performers.**
>
> WILL PENLAND, RETIRED SENIOR MANAGING DIRECTOR,
> HOUSTON OFFICE
> CB RICHARD ELLIS
> HOUSTON, TEXAS

Even though CBRE has been in the tenant representation business for years, they felt that the group from Trione & Gordon defined the leading edge of the profession. Thus the negotiation began between the small innovative company and the large, international player, both with the same aspiration in mind: To acquire a prominent position in an existing market category.

While this kind of transaction seems overwhelmingly commonsensical and appears clearly positive for both parties, such undertakings have an inordinately high failure rate. The goals are simple enough to define:

1. The nationally prominent firm wants to add substantial horsepower to its offerings by acquiring a small group of industry-leading high achievers.
2. The hard chargers, on the other hand, would like to see the capabilities they've developed applied on a national and international stage.

So what are the critical success factors in blending two culturally divergent organizations?

Most innovation occurs in small, high-energy, risk-accepting firms. But seldom do these creative pockets of professionals have the market muscle to amplify their successes to national or international scope. For that, a partnership with a larger group sometimes makes sense for both.

1 Budget enough time
Don't get in a hurry. The transition period for this kind of marriage should take at least two years.

2 Expect defectors
There will be people in both organizations who will find themselves uncomfortable with the new arrangements and resign. Typically, some from the acquiring company will resent what they see as special treatment of the newcomers while some of the newly acquired professionals will find the larger organization's administrative rigor to be strangling.

3 Expect a temporary loss of productivity
The new organizational realities are usually distracting for both groups and productivity can be expected to suffer.

4 Clearly define performance metrics for all parties
It is easy for some professionals on both sides to get caught up in the exhilarating feelings associated with a new organizational adventure. Management must communicate clear performance expectations for all.

According to Mr. Penland, "We've learned over the years that it's easy to acquire a new high-performance group and expect them to automatically align themselves with all of our processes. This almost always reduces their productivity to intolerably low levels. With the Trione & Gordon acquisition, we made the conscious decision to give them more latitude. After all, we wanted this deal because they're the best, so to immediately begin changing them made no sense to us."

> We helped define a new competitive category about 12 years ago based on a project delivery concept that became known as "alliancing." It involves the owner and contractor sharing construction risks. For a while, we were the experts and used this approach as a competitive advantage. Today, however, it has become more widely accepted, so we're busy designing the next competitive category.
>
> MIKE ROLLO, GENERAL MANAGER, BUSINESS SERVICES
> LEIGHTON CONTRACTORS
> SYDNEY, AUSTRALIA

5 Research and publish

This strategy is about learning high-value lessons that aren't common knowledge in your market segment and making certain all major buyers become aware. We believe this to be a particularly elegant way to go about the acquisition of prominence because it is good for your firm, good for the industry in which you are working, and can, when energetically executed be handsomely profitable.

The potency of this approach lies within the intersection of three concepts:

1 Heightened client intimacy

Learn more about your high-value clients than any of your competitors.

2 The empirical search for better solutions

Partner with your high-value clients to solve problems they consider critical.

3 Creative dissemination and branding of the results

Create, in the minds of your target audience, a clear relationship between your firm and unusually valuable solutions.

We see numerous examples of firms that, through diligent work, actually learn things that would put them at the forefront of their chosen market segments, but no one knows about it.

They don't put the word out typically for one of three reasons. (1) They see their hard-won knowledge as a competitive advantage and are afraid to give it away, or (2) they simply don't think about constantly and creatively communicating with a targeted audience, or (3) they've considered the issue and think that the communication process costs too much and they don't want to spend the money.

On the other hand, we see a wide variety of firms that understand the power of consistently delivering their message and their brand, but don't actually have anything interesting or insightful to say. Either way, this strategy fails.

Why giving away your knowledge is a good idea

Surveys tell us that many professional organizations are reticent to publish their most valuable knowledge sets for fear of giving away expensively acquired competitive advantages. Actually, we find just the opposite to be the case.

Here's what we've learned about the game:

Research and publishing is about learning high-value lessons that aren't common knowledge in your market segment and making certain all major buyers become aware. We believe this to be a particularly elegant way to go about the acquisition of prominence because it is good for your firm, good for the industry in which you are working and can, when energetically executed, be handsomely profitable.

1. Don't tell them everything … just enough to differentiate you from your competitors.
2. Your targeted audience won't read your material anyway. If you're lucky, they'll read the headlines. So make the headlines compelling.
3. If they find the material interesting, they won't attempt to act on it themselves, they'll call you for more information.
4. The publishing process is designed to maintain your firm in the conscious awareness of your targeted audience. The group that may actually pay more attention, however, is the general industry in which you function, making your firm appear to be a more prestigious place to work.

There is a fascinating lesson to learn from an international management consulting firm that published a new book about ten years ago. I would love to name names, but I fear doing so would be legally imprudent. This does not impair, however, the powerful lessons they learned about professional publishing.

The goal of their publishing effort was to establish themselves as the leading purveyor of a particular type of management consulting. Interestingly, they wanted to do more than tout the number of volumes sold. They wanted to be able to measure the impact their book had on the performance of the industry for which it was targeted. So they went to a great deal of expense to track not only sales, but who purchased the books and how the readers used the knowledge and techniques contained therein—a noble effort, indeed.

The results of their tracking were fascinating. Here's what they learned:

1. Total sales of 12,000 units, not bad for a professional book.
2. 78% of all buyers never opened the book.
3. 12% read only the front cover, back cover and table of contents.
4. 8% read as much as two chapters.
5. 2% of all buyers reported that they actually read the entire book.
6. Following the book publication, the firm's speaking engagements doubled.
7. New high-value employees, when asked what they knew about the firm, cited the book.

Lessons learned: Even though no one is actually is going to study your material, professional publishing is a powerful tool for acquiring prominence in any market category.

Fast Second: How Smart Companies Bypass Innovation to Enter and Dominate New Markets Constantinos Markides and Paul Geroski John Wiley & Sons, 2004

In 1946, two architecture professors and a graduate student—in their spare time—started a new practice that later became the single largest architectural firm in the world. William Caudill, John Rowlett, and Wally Scott found some unused office space over an old grocery store on University Avenue in College Station, Texas, and set up shop. Work didn't exactly pour in.

Since these men were an unusual admixture of academicians and practitioners, they were comfortable with the research and publishing strategy. The year was 1946, World War II was just over, and this young group of architects realized that the imminent return of hundreds of thousands of American soldiers would stimulate formation of an intense volume of new families (later to be referred to as 'baby boomers'). The offspring would soon easily overstress the nation's school system.

This new firm, CRS, consciously made the decision to employ the **research and publish strategy** in an attempt to capture the anticipated explosion in school design and construction.

Working with a wide variety of education experts, both researchers and practitioners, the architects began a rigorous program of trying to redefine the relationship between the learning process and the facilities in which it takes place.

It was interesting and relevant work, but the entire falderal would have come to nothing had they not prepared a series of research papers explaining and delineating their ideas and vigorously disseminated them to all relevant audiences in the field of public education. Keep in mind, this all occurred before CRS had designed a single school building.

> CRS had a unique approach that most firms do not follow today: publish everything—even your most innovative approaches that made you distinctive to clients.
>
> DR. ROBERT JOHNSON, EXECUTIVE DIRECTOR
> CRS CENTER, COLLEGE OF ARCHITECTURE
> TEXAS A&M UNIVERSITY, COLLEGE STATION, TEXAS

Within several years, CRS was designing more public education buildings than any firm in the US.

Lesson learned: They identified a specific market segment, differentiated themselves using the 'research and publishing strategy' and declared national ownership.

The logic of this approach is readily apparent, so why don't more firms do it? The answer most often provided: It takes too long and costs too much. So how have we seen companies shortcut the time and expense complaints?

1 Research without separate funding

Don't pay for the research by yourself. Work the empirical learning process into appropriate existing projects. If your ideas are sufficiently compelling, some clients will actually participate in extra funding. But even if they don't, most projects are rich sources for new learnings which can be documented with only modest extra effort.

2 Communication on a budget

Stay away from expensive consultants and paid space advertising. We've seen very effective communication programs managed and executed entirely in-house.

Everyone sends out the occasional brochure or direct mail piece; blogs and e-mail campaigns are becoming commonplace. The overwhelming deficiency we observe is that these efforts are seldom consistent. Lots of data exists that clearly explain: In the communication business, doggedly persistent repetition is more powerful than clever messages. These days, there are more inexpensive media options than ever before. Select the

Don't pay for the research by yourself.
Work the empirical learning process into appropriate existing projects.

6 Deeper specialization

Founded in 1955 in St. Louis, HOK is the largest architectural and engineering firm in the US and a leader in several market categories. "Today, our healthcare practice faces two distinct challenges," says Thomas H. Robson, Senior Vice President. "How do we maintain US leadership in a category that is becoming extremely crowded and, at the same time, export our experience and expertise to diverse cultures around the world?"

The strategy that HOK is using in the US healthcare market and one that often evolves in other segments is the process of deeper specialization. Twenty years ago, healthcare was its own separate market category demanding experts and special approaches. According to Robson, "Today we have distinct studios dedicated solely to research and teaching facilities, imaging centers, regional healthcare facilities, heart institutes, cancer treatment centers, and medical office buildings."

Robson's view is that, "The last generation could distinguish themselves in technical projects by being more inquisitive, asking more perceptive questions, and aiding the client in generating a better understanding of exactly how they wanted their new building to perform. Today, our clients expect us to be experts, to fully understand their facility development strategy, their business model, as well as their medical delivery processes. The value we are expected to add is not just about healthcare buildings, but how those buildings can facilitate healing processes as well as business processes."

Globalization

HOK is also a good place to look for lessons about the tricky process of capturing market leadership in other countries and cultures. Most global companies quickly learn that in dealing with overseas assignments, a cultural blending is going to be necessary. Says Robson, "The question quickly becomes: What part of our process can we export and what part must bend to local processes, customs and regulations through local delivery?"

He explains: "Two years ago, our London healthcare practice was struggling. We decided to try importing advanced planning and schematic design from the States while utilizing our existing London team for design development and CDs. We're now partnering with Skanska on two large hospitals in the UK. Utilizing our London experience as a template, we are proceeding similarly in India and Saudi Arabia."

The Quest for Global Dominance: Transforming Global Presence into Global Competitive Advantage
Vijay Govindarajan and Anil Gupta
John Wiley & Sons, 2001

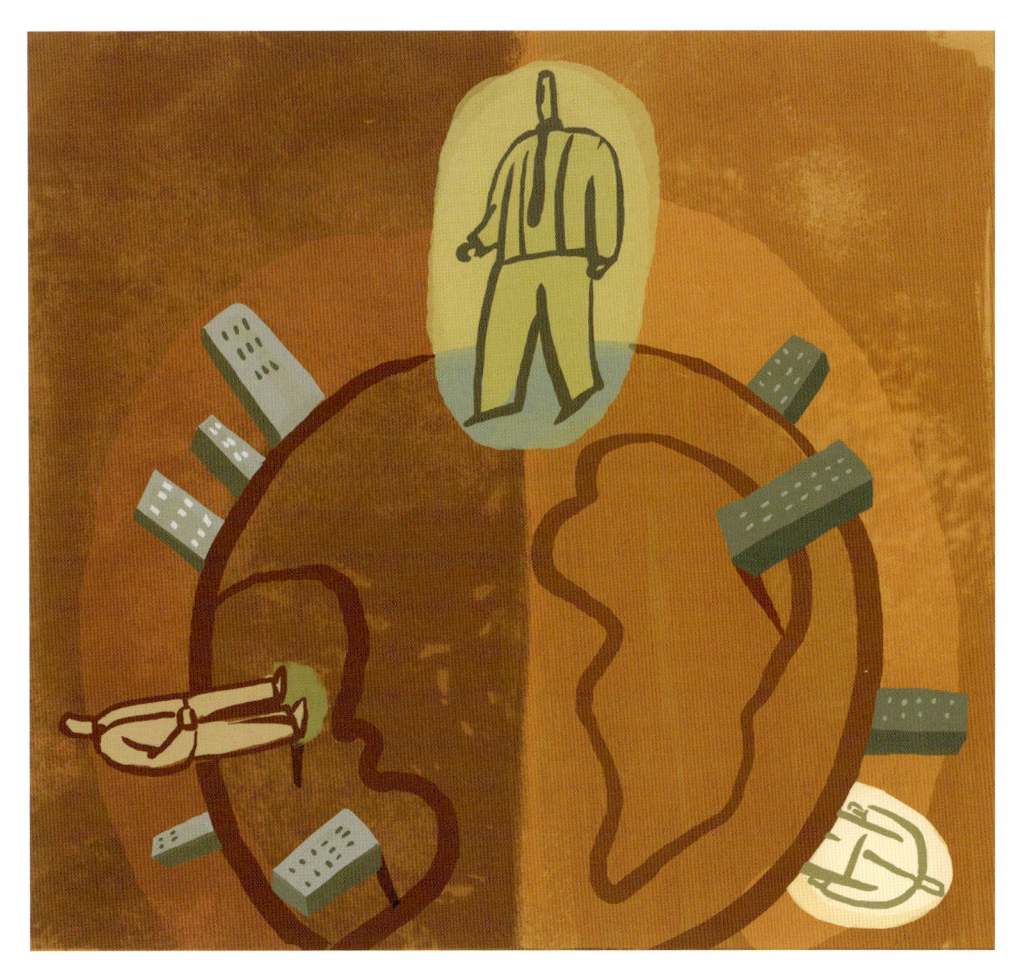

Most global companies quickly learn that in dealing with overseas assignments, a cultural blending is going to be necessary.

Paths that are overly worn

The following approaches have been widely used to the point of now becoming trite. Many of the last generation of market leaders were successful with these ideas, but overuse has left them a little worn around the edges. Tomorrow's prominent firms are going to have to focus and sharpen their approaches to respond to client demands for specialized expertise. Please don't write me to explain that you know of firms who still talk this way and are doing fine. I'm certain that's true, but it won't be the most effective path to future prominence.

> **Tired/overused/exhausted descriptions**
> Design excellence
> Technical excellence
> Creativity
> Professionalism
> Honesty/integrity
> Asking better questions

Is there anything wrong with excellence, creativity, professionalism, and honesty? Of course not. But firms relying on these descriptors to create differentiation will be disappointed.

We just discovered that my wife must have her hip replaced. I assure you that we're not interviewing highly respected general orthopedic surgeons who are honest, creative, and ask better questions. We want someone who replaces hips all day. There is a great deal of uncertainty in major surgery and we are not looking for creativity and honesty. We want to minimize risk. Most clients feel the same way about their new building projects.

Risk management

We haven't done this recently, but about 20 years ago we used to track the employment tenure for corporate managers who were put in charge of major building projects. Well over 50% were fired within one year of project completion. I assure you, all clients who interview architects, engineers, and builders are painfully aware of this phenomenon.

Most would readily swap a chance at cutting edge design for the increased possibility of keeping their jobs.

Risk management is paramount and most intuit that credentialed specialists are the safest way to go.

What role does customer intimacy play?

We've repeatedly discussed the idea, so much in fact that I'm certain you're bleary-eyed at seeing it again: The process of reshaping and redesigning market categories is one most effectively animated by heightened customer intimacy.

A deeper, richer, more lucrative relationship with your key clients sounds like a good thing, but few firms actually achieve it. Most are stuck with the standard business communication that comes as result of repeated meetings over breakfast, lunch, dinner, and the golf course.

> On many of our projects, we find that the client's senior management people, as the project progresses, distance themselves from the work and substitute subordinates to interact regularly with the design and building team. We work diligently to keep this from happening. It's impossible for us to establish the level of interaction we want if we let the real decision makers wander off.
>
> PETER ROGERS, DIRECTOR
> STANHOPE PLC
> LONDON, ENGLAND

So what techniques have we seen that work?
Interestingly, it's not about just doing your job well. You can win a new project, meet or exceed all client expectations, and still not develop a more perceptive understanding of the corporate strategies that gave rise to the new project that they hired you to execute. And that seems to be the point: It's not enough to understand that a client wants a new facility and to understand how they want that facility to perform and how much they want to spend and when they need it delivered. Those are the issues with which most architects, engineers and builders deal.

To uncover new, unmet, undescribed, unarticulated demand, AEC professionals must develop a robust appreciation of the intellectual and emotional machinations going on between the ears of those in corporate upper management.

> So what most of our clients are after are not property solutions, but business solutions of which property is an important part.
>
> NICK SHEPHERD, MANAGING PARTNER
> DRIVERS JONAS
> LONDON, ENGLAND

We began our investigation of this compelling issue by asking a collection of corporate upper management people: "When considering a new facility, we know that you describe to your design team what your company is about and what your goals are for the new building, but do you discuss the details of your most basic corporate strategies with your architects, engineers, or builders?" We got a resounding NO.

Why? The three reasons most often provided: (1) They wouldn't understand what I'm talking about, (2) they don't care because it's not necessary for them to understand the intricacies of our various strategies in order for them to design and build a new building for us, and (3) this information is proprietary and sensitive and there is no advantage to sharing it with anyone who can't materially help us execute it.

It's important that we distinguish the difference between a global explanation of corporate goals and an in-depth discourse on strategy. Corporations seem perfectly willing to explain their overall direction and conceptually how they intend to get there. But, these presentations are not where members of the AEC community are going to find new and powerful insights into unarticulated demand. The new ideas, the new approaches, the new high-value services are not going to grow out of information that is common knowledge. They will be born of the intimate knowledge that results from discussions and explorations of the details of senior management strategic thinking.

So, there you go.

We, therefore, have come to the conclusion that the most productive strategy for moving the entire AEC profession upstream is to overcome those three attitudes currently prevalent in corporate upper management.

It's critical that we take a moment to let that information soak in. Let's recap. When the topic is "business strategies that motivate building strategies," many corporate executive decision makers think that AEC professionals:

1. Don't understand
2. Don't care
3. Can't help

They think that, with us, it's all about buildings.

The new ideas, the new approaches, the new high-value services are not going to grow out of information that is common knowledge. They will be born of the intimate knowledge that results from discussions and explorations of the details of senior management strategic thinking.

Category Ownership

How will the next generation develop and exploit new categories?

> **Good companies will meet needs; great companies will create markets.**
>
> PHILIP KOTLER
> IN *MARKETING MANAGEMENT: ANALYSIS, PLANNING, IMPLEMENTATION AND CONTROL*
> PRENTICE HALL, 1996

As you know by now, we believe the single most powerful strategy for creating revised market categories will be through a partnership with your best clients. They will lead you to the development of higher value approaches but only if you know how to engage the right people on the right issues.

Opportunity mining

The issue here is pretty simple. We are certain that the path to higher value, more prestige, and more revenue for the AEC profession goes right through the practice of increased customer intimacy. We are going to get there by asking better questions of the right people and being willing to reshape our offerings to respond to, what today, we refer to as unarticulated requirements.

We can be more valuable to the corporate world than they currently know and, I believe, more than we currently know.

So what do we do? How will the next generation of market leaders overcome the three objections delineated earlier?

Learn
Learn their industry, their language and their culture.

Access
In-depth discussions with the wrong people are useless. You must gain access to senior management decision makers.

Engage
Using their language and their culture, engage them in conversations about the detailed business conundrums in which they find themselves. In order to this, we must demonstrate dexterity with the issues they find most compelling. A blank stare terminates the process. We must fully understand what they are facing and communicate with them in their chosen vernacular.

As you know by now, we believe the single most powerful strategy for creating revised market categories will be through a partnership with your best clients. They will lead you to the development of higher-value approaches but only if you know how to engage the right people on the right issues.

Interpret

These conversations, when carefully dissected, will contain seedling ideas that have the capacity to grow into more valuable services. We are engaging senior management in explorations of business strategy, not buildings. And yet, practically all explorations of strategy disclose impacts on the built environment.

Rethink

Armed with a new and deeper perception of our best client's struggles, we should be able to design approaches that describe how our knowledge of design and building can actually be part of their strategic business solutions.

Propose

It's now time to make our case. It should begin with a clear restatement of the strategic business issues that we understand our client to be facing, once again, explained in their jargon. Next, we describe how the built environment is one of the weapons that senior management can utilize in attacking their strategic business opportunities.

Demonstrate

Sooner of later, someone is going to give us the chance to perform. And when that happens, we must document our process and all relevant outcomes. Creative business approaches can't be taken seriously without objective metrics to support them.

Brand

When successful new service extensions occur and market category definitions are effectively redefined, we must immediately brand our victories. That means naming the new approach, perhaps with a registered trademark, and making certain everyone in our targeted audience knows about it.

What on planet Earth could possibly be more fun than reshaping an existing market segment, declaring ownership, and utilizing the process to create a new national or international presence?

Persistent Branding
Whether you like it or not, your company has a brand. It has created a complex amalgam of impressions on the constituencies with which it deals: past clients, current clients, current employees, potential employees, industry leaders, community leaders and, actually, everyone with whom it interacts.

Chapter 5

PERSISTENT BRANDING

What's so important about a brand?

How do today's AEC market leaders utilize branding?

What role does competitive intelligence play in branding?

How will the next generation of leaders brand?

Your brand

Whether you like it or not, your company has a brand. It has created a complex amalgam of impressions on the constituencies with which it deals: past clients, current clients, current employees, potential employees, industry leaders, community leaders. Actually, everyone with whom it interacts. The branding phenomenon, therefore, is guaranteed to happen, so the only question we face in the AEC industry is whether to let it occur haphazardly or to manage the process purposefully.

Branding is not a new concept. It has been studied extensively for 200 years, particularly among companies that design, manufacture, and sell consumer products. As a matter of fact, the business literature is covered up in titles about branding. One publishing source estimates that over 12,000 titles are currently in print explaining the infinite strategies available for exploiting the visceral excitement of powerful brands.

> I think there's a broad misunderstanding in the design community as to exactly what a brand is. Most tend to think of it as just being reputation.
>
> TIM BLACK, DIRECTOR
> BKK ARCHITECTS
> MELBOURNE, AUSTRALIA

Revised thinking

For the past 20 years, we have spent a great deal of energy thinking about how best AEC firms can create and maintain effective company brands. Recently, however, we see the game changing and no longer think in those terms. Two factors have greatly impacted our views:

1 Market segment specialization

We're all aware that major project buyers no longer search for good, well-regarded architects, engineers, and builders. They want specialists—companies and individuals that have amassed substantial expertise in precisely the kind of project the client is trying to create. In these waters, the traditional generalist is at a substantial disadvantage. Firms have responded by creating groups of specialists within their larger organization who can respond to these demands for in-depth expertise.

2 Globalization

Today, many firms are required to perform on two stages at the same time: They must present themselves as extensively in reach with access to global resources and, at the same time, demonstrate dexterity with local knowledge, contacts, customs, and regulations.

Two levels of branding

These realities give rise to the necessity of two separate but equally important branding programs:

1 Company level branding
2 Service level branding

Company level	Service level
Public image	Brand management per market category
Reputation	Clear, consistent, repetitive messages
Stature	describing the competitive value
Prominence	advantage for each service line

"Although we compete in a wide variety of market segments, we want Golder to present a consistent face to the clients served by each of our 120 offices," says Frederick Firlotte, President and CEO of Golder Associates in Montreal. This far-flung engineering giant currently operates in 30 countries and must deal with 14 different languages. Marketing and branding strategies obviously vary somewhat depending on locale, but Golder utilizes a comprehensive global tag line: "We're engineering the earth's development while preserving the earth's integrity."

Kellogg on Branding: The Marketing Faculty of the Kellogg School of Management
Philip Kotler and Alice Tybout
John Wiley & Sons, 2005

Terminology
To get the branding conversation started, here is a brief dictionary of generally accepted branding terms and how we see their relevance in the AEC industry:

Brand: the sum total of impressions about your company held by any defined audience
Among companies in the AEC arena, the term "brand" is often used interchangeably with "image" or "reputation", or "professional stature" or "industry prominence." While these terms are interrelated, we don't see them as synonymous. Our concern is that while reputation, stature, and prominence are concepts to be admired and valued, they connote a sense of passivity—they just happen as a natural result of properly taking care of business over a long period of time. The idea of "branding," however, suggests a series of steps that we execute with purpose: Something to be actively designed, communicated, monitored, and managed.

Brand alignment: the practice of linking your brand strategy to your company's vision, values, marketing plan, and business plan
Our experience is that while most branding programs in the AEC industry are generally anemic, other planning processes are alive and well. Although the smallest firms among us typically don't formally plan, everyone else does. It is unusual not to find a company with a competently crafted vision statement, values statement, marketing plan, and general growth strategy. Branding, on the other hand, is often assumed to occur of its own free will.

Brand credibility: a measure of how well your company actually delivers the commitments inherent in your brand message
Companies that don't regularly deliver the outcomes promised by their brand will quickly see it relegated to the *Encyclopedia of Hollow Slogans*. In our industry, however, this is rarely an issue. Successful firms are traditionally built on years of competent professional service.

Our problem is not that we don't deliver on the promises of our brands; the problem is that we don't have well-developed brands to begin with.

Brand equity: the worth that a brand supplies your company beyond its intrinsic value

Imagine the last successful competitive presentation you made, hoping to acquire a new project. Now imagine you work for a new, unknown company and you make the exact same presentation. Would you still have gotten the work? If so, your brand didn't do you any good and therefore has no equity.

Several studies have recently looked at the relationship between reputation and stock price. Although this process requires sifting through years of complex data and searching for statistical relationships which must be framed as tentative at best, we all intuit in the AEC world that professional stature, or a company brand, is worth money in the bank.

Marketplace Masters: How Professional Service Firms Compete to Win
Suzanne C. Lowe
Praeger Publishers, 2004

Brand equity: the worth that a brand supplies your company beyond its intrinsic value.

Brand essence: the simplest expression of your brand's promise
Humans don't process complexity very well and therefore can't be expected to respond to any message that isn't clear, simply stated, and emotionally appealing. If you will notice, in the last statement there was absolutely no mention of intellectual content. The scientific community assures us that *Homo sapiens* possess a prefrontal cortex, the right hemisphere of which produces rational thought. I see no evidence of any such phenomenon.

When considering the essence of your brand's message, forget the intellectual stuff and focus on emotion.

What's the difference between a brand that actually possesses a relevant essence and one that has been decorated with a clever but hollow slogan? *Meaning*. The test is incredibly simple. Write the essence of your brand on a piece of paper, show it to audiences you care about, and ask them what it means. If they can't answer or everyone gives you a different response, you don't have a brand. You have a slogan. On the other hand, if the words on the paper produce actual meaning in the minds of the reader, you have created something of value.

Brand experience: the sum of all interactions responsible for communicating your brand message
Too many of our research partners see branding as a process of authoring a couple of clever phrases and plastering them on business cards, brochures, and web sites. Some are even conscientious about making certain all of the firm's printed material is consistently designed. This is actually a good idea, but it doesn't have much to do with branding. For firms in the AEC industry, the message of who we are gets integrated into the world by way of human interaction, not printed material.

Training your people to deliver your brand through their behavior is far more powerful than slogan dissemination.

The following messages are from some of the world's leading AEC firms. Are they the essence of powerful brands or are they slogans whose job it is simply to adorn business cards?

The Right People. The Right Results.
Hunt Construction Group

A World of Solutions
The Shaw Group

What an Architect Should Be
FKP Architects

People. Building. Progress.
Stuart Olson

Architect of Ideas
Gensler

Ideas Work
HOK

Building Success Stories
Brasfield & Gorrie

Ideas Change Everything
NBBJ Architects

Expertise Imagination Partnership Performance
Cannon Design

Integrity Commitment Teamwork
Turner Construction

Global Issues Local Solutions
Golder Associates

Consulting + Engineering + Technology + Construction
Syska + Hennessy

Expect More
Drivers Jonas

For firms in the AEC industry, the message of who we are gets integrated into the world by way of human interaction, not printed material.

Brand extension: the process of taking an existing set of positive brand attributes and applying those same characteristics to a different set of services

This particular strategy is extremely active right now as all firms scramble for a sustainable competitive advantage. We see many groups well regarded in their industry adding a wide variety of services from other industries to their traditional offerings.

Real estate consultants, architects, engineers, and builders are aggressively attempting brand extensions by integrating new services such as project management, management consulting, facility management, financial strategies, facility maintenance, and equipment consulting. As we all know, some brand extension programs are wildly successful and others fall flat.

Brand gap: the difference between the commitments inherent in your brand's message and your company's actual performance

Performance promises in our industry are typically not made through our brands—they're made in the presentation, negotiation and contract execution process. If a gap occurs, it lives between what we said we could do in the early stages of the project and what ended up happening. These disappointing gaps occur for an endless variety of reasons, some under our control and some not. Nobody likes them and all professionals take them very seriously. The most prominent AEC firms are experts at **expectation management**.

Brand metrics: analytics that allow you to measure the effectiveness and value of your brand

Branding is one of those topics that is easy to discuss and hard to measure. We have witnessed many AEC firms that intuit the value of a strong brand, and then abandon their efforts simply because there was no objective way to measure results. It's easy to track the money we spend on branding so if specific financial impacts are not readily apparent, most programs get shut off. Objectively measuring the value of your brand along with its financial impact is perfectly doable and we'll discuss how later in this chapter.

Brand pushback: a measure of resistance to your message by any particular audience

This usually occurs when your company promulgates a new message that doesn't coincide with what your target audience already knows, or thinks they know, about you. They simply don't believe it and most likely will respond by ignoring the new message. We have seen architectural firms, engineering firms, and construction firms all attempt brand extension programs into such things as management consulting, financial services, and real estate consulting. As a rule, these attempts at extension are rationally based since the new services clearly relate to the firm's existing services and brands. Invariably, these programs will be met with some "brand pushback." This is not necessarily negative, just part of the process of extension. Properly managed, it can be overcome. Improperly managed, it will kill the new extension effort.

Designing Brand Identity: A Complete Guide to Creating, Building, and Maintaining Strong Brands, 2nd Edition
Alina Wheeler
Jossey-Bass, 2006

Brand strategy: a comprehensive program to design your message, align it with business strategy and communicate it to targeted audiences

Having worked with leading AEC firms for years on the topic of competitiveness, we clearly find that branding efforts typically lag far behind other industries. We are officially tired of speculating about why this happens. It doesn't make any difference. Any annual business plan that does not include an integrated branding strategy will simply not serve to effectively position your firm in the hearts and minds of targeted buyers.

Brand touchpoints: an inventory of all interaction between your company and its targeted constituents

We find that most AEC firms actually have more touchpoints than they are aware of. Fortunately, or unfortunately, they all count and work together to create your brand experience. When first considering a brand strategy, most firms execute an inventory of all touchpoints, invariably an interesting and instructive exercise.

More powerful lessons are learned, however, when an inventory is taken of how all the touchpoints touch each other.

Any annual business plan that does not include an integrated branding strategy will simply not serve to effectively position your firm in the hearts and minds of targeted buyers.

98   The New Competitiveness

You will find that they are inextricably networked. We have many examples of how mistreating an apparently minor contact can have unexpectedly painful results in larger arenas.

What's so important about a brand?

When competing for work, we must invariably navigate several predictable stages.

1. We look for leads.
2. We attempt to make contact with the identified prospect.
3. We send designed material to get us on the "short list."
4. We compete in the formal presentation and proposal process.

Each step along the way becomes measurably easier if we have a strong brand preceding our marketing actions. It is not a suitable use of our limited intellects to further belabor a point that we all intuit; a strong brand is worth money.

> One of the things you want a brand to do as a professional services firm is to open the door for you, so by the time I get in front of a prospective client, our brand has already done half the job.
>
> GREGORY HODKINSON, CHAIRMAN
> ARUP AMERICAS
> NEW YORK, NEW YORK

How do today's AEC market leaders utilize branding?

Here, we have good news and bad news. If we were teaching a class on the persuasive influence of branding, we wouldn't use many examples from the AEC community. Our experience is that only the largest and most sophisticated firms produce any sort of serious attempt at branding their services. The bad news is that this means many competent, well-regarded companies are leaving some powerful competitive weapons sitting around unused.

The good news is that if you have a bright, young venture and want to slam it into fast forward, your erstwhile competitors seem willing to step aside and give you unfettered access to branding, a competitive implement shown to work handsomely in other industries.

Arup

Arup is an international engineering and consulting firm with offices in 86 cities. Gregory Hodkinson, Chairman of Arup Americas, explains: "Because our company has grown inside a professional engineering business culture, it has traditionally been considered unacceptable to regard ourselves as a typical commercial venture. Thirty years ago, we would never outwardly promote ourselves or our work. As a young engineer, the first Arup project I got to work on was the Sydney Opera House. Although it was a new project with worldwide prestige, we made no announcement of being selected for the work. We simply executed the project to the best of our ability and waited for other commissions to walk in the door. Needless to say all of that has changed. We now recognize the immense value of our international brand and that it must be nurtured and managed."

The most effective efforts we have witnessed are by firms who understand that they serve different constituencies, each of which has its own set of interests. Any successful branding program must address these divergent groups with their own independent message.

And even though each constituency must be dealt with individually, it is important to tie it all together with a company-wide image.

Stuart Olson

Stuart Olson, founded 38 years ago, is now one of Western Canada's largest construction companies. About 18 months ago, they began executing a new branding program in response to two very different challenges: (1) a transition from family to public ownership, and (2) extensive growth necessitating the addition of new employees at all levels of the firm.

According to Al Stowkowy, President and COO, "Our company fills different needs for different constituencies, so one message hardly seemed appropriate." Olson defined six groups it serves as worthy of specialized communication and set about to determine exactly how to describe the desired relationship.

The most effective efforts we have witnessed are by firms who understand that they serve different constituencies, each of which has its own set of interests. And even though each constituency must be dealt with individually, it is important to tie it all together with a company-wide image.

> **Stuart Olson constituents and the message designed for each:**
> - **Clients:** You trust us because we keep our promises. We work hard to understand your business and what you want so that we will exceed your expectations.
> - **Consultants:** You have confidence in us because we understand and respect your design. In partnership, we work together to exceed clients' expectations.
> - **Trade contractors:** Our relationship with you is one of absolute fairness and respect. Our project leadership style establishes an environment for success—a place in which you will thrive.
> - **Employees:** We are open and honest with each other. Together, we grow challenging careers that balance work and leisure in an organization that rewards excellence.
> - **Shareholders:** We do not compromise on value or values. Shareholders consistently find us accountable, profitable, and ethical.
> - **Community:** We support people where we live. We are responsible to our community since we live here too.
> - **Environment:** We will lead in green construction because we believe in it. We strive to protect the environment on every job.

Having addressed each of its defined interest groups, Olson ties it all together with a company-wide line that they believe to be the essence of their branding efforts: *People. Building. Progress.*

This effort is now 18 months old and senior management is pleased with the results. They believe they now have more consistent visibility in the marketplace and that their constituents see them more positively. Says Stowkowy, "Our volume literally doubled last year. No doubt the healthy economy is part of that, but it also demonstrates the success of our attention to branding."

What role does competitive intelligence play in branding?

When we look at the branding efforts of today's leading firms, what lessons can we take away? There are obviously many ways to analyze what is and what is not working, but there are two indicators that we believe to be the most powerful predictors of branding success.

1. Branding strategy foundation: What assumptions support the branding strategy?
2. Branding program execution: How much organizational energy has been assigned to the branding program?

The vast majority of strategies we reviewed were based on some complex bricolage of information composed equally of empirics, superstition, hearsay, gossip, chit-chat, and rumor.

The assumptions were typically intuitive and unevenly applied, resulting in information unworthy of being the basis of an actionable strategy.

We believe that there are only two categories of information worthy of being the basis of a productive and vigorous branding campaign:

1. Objectively collected market intelligence
2. Individually perceptive observations born of enhanced customer intimacy

Please notice that both of the above listed intelligence-gathering practices are externally focused. There is no mention of the senior management team getting together in a conference room and, armed with two pots of coffee in a room full of prejudice, imagining what current dynamics are driving relevant market forces. Branding strategies based on personal opinion, no matter how experienced and well described those opinions might be, are doomed to look and perform like everyone else's.

High-performance branding must be based on data not generally available to all competitors.

Competitive intelligence
So, how are today's market leaders utilizing competitive intelligence to underpin their branding efforts? It is not a startling idea that every company functioning in a highly competitive arena should attempt to understand the appurtenant rules of engagement. During the course of our research, we found absolutely no firms, large or small, that didn't claim to have a working knowledge of all aspects of the competitive landscape in which they function. Not surprising.

Intelligence-gathering practices are externally focused. This does not involve the senior management team getting together in a conference room and, armed with two pots of coffee in a room full of prejudice, imagining what current dynamics are driving relevant market forces. Branding strategies based on personal opinion, no matter how experienced and well described those opinions might be, are doomed to look and perform like everyone else's.

Here's what is: Since every company knows this is important stuff, and every company is actively interested in good competitive intelligence, why, then, are most of us so bad at it?

> **CI** is actually the legal and ethical process of collecting, analyzing, and applying information about the capabilities, vulnerabilities, and intentions of the competition.
>
> STEPHEN H. MILLER, SPOKESMAN
> SOCIETY OF COMPETITIVE INTELLIGENCE PROFESSIONALS

Competitive Intelligence
Jim Underwood
John Wiley & Sons, 2002

Better decisions

Competitive intelligence programs are about making better decisions, an issue with which management thinkers have been grappling for years. Why do some executive teams repeatedly outperform others? Why do some always seem to be in the finals for major projects, are better at applying new technologies, can repeatedly commercialize new approaches, and always seem to be one step ahead of their worthy competitors?

Competitive Intelligence
Since every company knows competitive intelligence is important stuff, and every company is actively interested in good competitive intelligence, why, then, are most of us so bad at it?

Persistent Branding

Are they more intelligent? More educated? More creative? More aggressive? Better organized?

No.

We have spent years observing AEC executives while they wrestle with the complexities of planning and executing competitive branding strategies, and two variables jump to the top of the page when we compare success rates:

Perceptive questions

Market-leading executives ask more relevant questions about three topics: (1) the behavior of the markets in which they compete, (2) the behavior of the potential clients they target, and (3) the behavior of the companies with which they must contend.

Quality data

They gather data that gets to the heart of questions they ask, not some wildly diverse assemblage of opinions, views, and war stories.

On the other hand, we observe on a daily basis companies that ask the same old questions that their competitors ask and gather the same old data that their competitors gather.

It's no wonder the competitive strategies of most firms look remarkably similar.

> If you are ignorant of both your enemy and yourself, you are certain to be defeated in every battle. If you know yourself, but not your enemy, for every battle won, you will suffer a loss. If you know your enemy and yourself, you will win every battle.
>
> ---
> SUN TZU
>
> THE ART OF WAR
>
> C. 500 BCE

The Secret Language of Competitive Intelligence
Leonard Fuld
Crown Business, 2006

When working with current market leaders on their competitive intelligence efforts, we quickly learned that only the more discriminating companies realize they have holes in their understanding of the competitive environs.

> **W**e're always trying to improve our understanding of the competitive landscape. We think we know our competitors and their products, but I'd like to know more about how they work with their clients and what they do differently or better or worse than us.
>
> ALEC TZANNES, PRESIDENT
> ROYAL AUSTRALIAN INSTITUTE OF ARCHITECTS 2007
> SYDNEY, AUSTRALIA

The frightening executives are the ones that imagine their intuitive and incomplete data sets are perfectly capable of being the basis for successful future strategies.

We found that today's AEC market leaders used their competitive intelligence energies to generate information in four areas:

1 *Competitors*
- Who are they?
- What services do they offer?
- What are their fees?
- How do our services and fees compare?

2 *Potential clients*
- Who, in our market area, has the capacity to buy our services?
- What are their typical selection criteria?
- Who specifically, within our target companies, makes the decisions?
- What projects do they have coming up?
- How do we get on the short list?

3 *Centers of influence*
- Who impacts or influences our target clients' decisions?
- How do we get our message in front of the influencers?

4 *Competitive selling techniques*
- How do our competitors present themselves?
- What do their proposals look like?
- What selling techniques do our target clients respond to?
- How do our selling techniques compare to our competitors?

> **W**e are effective competitors and have developed over the years an intuitive understanding of who our competitors are, what our potential clients are about, and what our differentiators are, but I worry that some of our information is fanciful.
>
> SCOTT REED, WESTERN REGION MARKET & PRACTICE LEADER
> CANNON DESIGN
> LOS ANGELES, CALIFORNIA

The above questions all make perfect sense. The problem is that most current market leaders don't display any particularly organized methods for gathering the information or analyzing it with much purpose.

So how do these companies pursue this information?

Today, competent CI programs don't require dramatic Boris and Natasha, spy-type behavior, and dumpster diving is no longer required.

Ninety-eight percent of everything you need to know is publicly available, inexpensive, and there for the taking.

Overwhelmingly, today's market leaders pursue the information they want through personal networking. Successful firms develop, over the years, a complex set of personal contact tentacles that weave their way throughout the business community.

Any professional with a robust personal network in today's world is a major asset. Many firms, in fact, pay quite a premium to those willing to spend their days and nights networking decision makers. The tools of the trade are usually pricy, but one major project heals all the pain.

However, it's not the only way to proceed. Although personal networking will never be replaced, the next generation of market leaders will have to supplement the hand shaking and cocktail parties with well-designed programs that allow them to know more than their adversaries about effective competition.

Professional Journal
The Competitive Intelligence Review
The Society of Competitive Intelligence Professionals

Overwhelmingly, today's market leaders pursue the information they want through personal networking. Successful firms develop, over the years, a complex set of personal contact tentacles that weave their way throughout the business community.

> Questions about the competitive landscape are very interesting. We're reasonably well up to speed on our competitors and their products, and it always helps to have specific information about our potential clients.
>
> ANDREW MORRIS, SENIOR DIRECTOR
> ROGERS STIRK HARBOUR & PARTNERS
> LONDON, ENGLAND

How will the next generation of leaders brand?

We see the next generation of market leaders attacking the issues of professional prominence, stature, reputation, and branding with more focus and energy than is the case today. Specifically, there are five areas that will be addressed:

1 Purposeful versus intuitive
Today's leaders substantially overuse their intuitions when approaching branding. Although this generation seems to think this is important stuff, too many of them are feeling instead of thinking their way through the process. Future market leaders will have purposeful branding campaigns based on competitive intelligence that is more perceptive than the competition's.

2 Alignment with business strategy
Today's successful firms all produce materials that make up what they call their business strategy. Unfortunately, they spend way too much time budgeting and way too little time actually producing an active and vital competitive strategy. This approach typically produces a collection of initiatives that are individually interesting but collectively don't create a coherent, integrated path to increased competitiveness. The next generation won't consider their business strategies complete without a fully integrated program to creatively brand not only their company's image, but each individual service they provide.

3 Better competitive intelligence
Any powerful competitive strategy is an amalgam of moving parts, hopefully all working together to produce measurable results. At the heart of this machine lives a perceptive understanding of the competitive landscape, an understanding that penetrates the typical barriers to competitive intelligence. Here, intuitions are an effective starting place, but they won't get you where you need to be.

Even though most AEC firms are fully aware that critical data is not just going to find its way into their companies, very few have anything that approaches an organized competitive intelligence program.

And yet we observe that most would like to put something in place that at least gets them started.

Today's successful firms all produce materials that make up what they call their business strategy. Unfortunately, they spend way too much time budgeting and way too little time actually producing an active and vital competitive strategy.

Here's what we observe the important initial steps to be:

- Begin with specific questions required to formalize a competitive branding strategy and don't gather data that doesn't specifically add to the picture.
- Appoint a chief intelligence officer who is a well-respected person from senior management.
- Assign staff with hours allocated to gathering intelligence.
- Create a specific competitive intelligence mission statement.
- Create a specific ethics statement.
- Educate your employees about gathering intelligence. Get everyone involved and publicly recognize those who excel.
- Create a central competitive intelligence file.
- Create a small team to interpret the data and make recommendations.
- Hold regularly scheduled competitive intelligence presentations.
- Develop performance metrics. (1) Track the acquisition of new leads: How many? Who generated them? Source? (2) Track leads that produce presentations. (3) Track sales win and loss record.
- Publicly reward success.

4 Objectively measure outcomes

The study of competitiveness is fascinating, but it seems to provide a collection of "rules of thumb" rather than a set of "immutable laws." But, here's one:

Companies that don't find a way to attach some measure of financial impact to their CI efforts will sooner or later abandon them.

Brand Innovation Manifesto: How to Build Brands, Redefine Markets and Defy Conventions
John T. Grant
John Wiley & Sons, 2006

5 Routine review and adjustment

It usually doesn't take long for a new competitive intelligence program to begin screaming its results. Some activities will be richly productive, others will be useless and it won't take a complex analysis to figure it out. Regularly scheduled evaluation is a must.

Getting a branding program started

Here are the steps we've identified as places to start:

Step 1: Create an internal branding team

This group should be small, led by a senior executive, and contain members from a cross-section of the company. The larger companies will probably utilize one or more consultants for this undertaking while the smaller ones may attempt to do this in-house. Either will work, but the following elements are critical. Omission of any one of them assures ignominious defeat:

(a) *Draft a mission statement.* This group will be responsible for generating a compelling branding message for each of the project categories in which you compete. They will also address the issue of professional stature for the firm as a whole.
(b) *Create an operating budget.* If there is no operating budget and specific time allotted, after the first several meetings, the entire team will be consumed by their other project work and the branding team will disintegrate.
(c) *Develop methods for measuring their achievements.* By now, you must feel inundated by the repeated plea for measurement.

We all know that you don't get the behavior you talk about, you get the behavior you measure.

I'll try not to mention this more than a couple more times.
(e) *Develop a system to reward accomplishment.* It is far more important that the reward system be public than monetary.

Step 2: Measure the impact of your existing brands

How are our existing brands currently perceived by the critical audiences that matter to us?

The vital item here is the use of objective data that provides an accurate picture of where you stand relative to your competitors in each pertinent project category. We have witnessed too many firms that believe they already know this information. Typically, over a cup of coffee, the firm leaders gather, glance skyward, and describe how they stack up against each of their competitors. This exercise, while charming and often replete with a collection of colorful anecdotes, is not only useless but actually damaging. It is fraught with guesses, prejudices, half-truths, fables, and folklore. If your company chooses to hold one of these séances, don't miss it, it will be perfectly entertaining. But don't dare base any branding decisions on what you hear.

For branding efforts to bear fruit, they must be based on objectively acquired information about your own performance, your competitors' performance and how your potential clients perceive the difference. Does the information you gather need to be objectively unassailable? No. It doesn't even need to be true. This is about human perception.

If you would like to witness masters in action, simply turn on your TV for a few minutes and watch our duly elected political titans manipulate public perception while keeping the truth at a safe distance.

Design relevant survey techniques. We find companies utilizing a wide variety of information-gathering techniques: personal interviews, focus groups, mail surveys, web surveys, etc. Obviously they each have their strong suits and perhaps a mixture works best. This process is indispensable because it provides the foundation for all subsequent branding and marketing decisions.

Design relevant survey techniques. We find companies utilizing a wide variety of information-gathering techniques: personal interviews, focus groups, mail surveys, web surveys, etc. Obviously they each have their strong suits and perhaps a mixture works best. This process is indispensable because it provides the foundation for all subsequent branding and marketing decisions. Inaccurate, incomplete, or irrelevant data will render the entire process harmfully moot.

We have also seen this process executed successfully by in-house employees as well as credentialed consultants and, of course, combinations of the two.

A wide variety of advisors operate in this space: marketing consultants, market research firms, advertising firms, PR firms, etc. Be careful with these people, they can be expensive, but if they actually generate actionable data, it should be well worth the price. And look for ways to compensate them based on performance. I've never liked the idea that these people usually get their full retail fee even if their work completely misses the mark.

> We have seen many substantial companies track the performance and pricing of their largest and most aggressive competitors. This clearly makes sense but may cause you to miss game-changing indicators. It's important to identify at least two bright, young hard-charging companies to monitor. They are often more fruitful harbingers of market movement than larger players.

Step 3: Execute a branding identity investigation
For each market segment in which you function, your services must be identified with a meaningful and engaging message that distinguishes you from those low-grade, dull plodders with whom you compete. The thought process is certainly simple enough, but

we continue to gather a long list of companies whose branding approaches are not only uninspired, but banal in the extreme.

There are a few simple questions to answer:

1. In which market categories do we compete?
2. Who are the major buyers?
3. What do they want from us?
4. How are our competitors presenting themselves in these categories?
5. How do our targeted buyers perceive us relative to the competitive field?

Simple enough. So why do so many good companies proceed with such a painful lack of inspiration and creativity? Here's the most quizzical observation of all: Architectural firms that design the most prominently inventive buildings often market themselves, for all practical purposes, in secret. Engineering firms that regularly solve incredibly complex building problems with creativity and efficiency present themselves to the marketplace with the inspiration of funeral directors. And large, credentialed builders who can construct anything that architects and engineers can imagine, and by the way, do it on time and on budget, regularly present a face to the industry that is as bland as day-old oatmeal.

Why don't these energetic, impressive companies create brands that match their professionalism, focus, and performance?

Step 4: Define the perception delta
At this point, the branding team should have a clear, objective understanding of how each of your services is perceived and a similarly clear understanding of how you would like them to be perceived. So, why the delta?

Is it because you just haven't gotten the word out to the right people about who you really are? Or is the perception in the marketplace reasonably accurate and you feel your company simply needs to make it more clear?

Step 5: Devise branding action strategy
At this point, the branding team knows how each brand is perceived and how they want it to be perceived. It's time to get the message out.

The most effective propagation techniques depend on the message and the audience. There are hundreds of books describing your options as well as a plethora of consultants on which to call.

Your branding program isn't complete until the branding team knows how each brand is perceived and how they want it to be perceived. Then, it's time to get the message out.

We've learned the following lessons concerning dissemination from the world's most aggressive branders:

1 *Get started even if your budget is modest.* Expensive communication techniques such as extensive full-color brochures, highly-produced web sites, radio, TV, and paid space advertising aren't necessarily the most effective ways to reach out.

Persistent Branding 117

Interact with your target audiences with a scalpel, not a meat ax.

2   *Your single most efficacious communication practice will be the utilization of your staff.* They are the ones who are responsible for delivering your brand promise anyway, so it makes sense to fully train them on how important your brand is and how you want it delivered.

<small>The Expressive Organization: Linking Identity, Reputation, and the Corporate Brand
Schulz, Hatch, Larsen
Oxford University Press, 2002</small>

Syska + Hennessy, an engineering and construction firm headquartered in New York, has 14 offices with a staff of 650. They cleverly engage their people in distributing the essence of their brand by actively promoting the idea of "micro-branding." William Caretsky, Senior Vice President, explains, "We don't think that branding messages aimed broadly at the public gets us much. We know who we want to talk to and we know that personal attention from one of our principals is far more productive than brochures or advertisements or web pages. Our most effective communication will always come from our people."

> At Cannon Design, the first target for our branding message is our own people. We realize that, in the end, they will be the most powerful conduit we have for delivering our brand to the audiences we target.
>
> LEE BRENNAN, PRINCIPAL
> CANNON DESIGN
> LOS ANGELES, CALIFORNIA

Step 6: Evaluate and adjust
Branding strategies need to be as vibrant and variable as the market forces that drive them. The companies that get the most value from their branding efforts realize that they never get it right.

Every program is a test that must be evaluated and all branding activities are seen, from their inception, as temporary.

Summary

For those of you who fell asleep during this chapter but would still like something to take away:

1. Branding is critical. Don't do it intuitively.
2. It must fit, hand-in-glove, with the rest of your business strategy.
3. All branding programs must be based on superior competitive intelligence.
4. Branding efforts must address two distinct spheres at the same time: (a) total firm public image and (b) individual service brands.
5. Measure the outcomes of your branding program.
6. Publicly reward achievement.

Marketing Breakthroughs

The perception of a stable market segment is an illusion. Every market category is in the process of expanding or contracting at all times. And any company that sees their revenue as stable over the years is either grabbing a larger share of a shrinking segment or getting less and less of an expanding one.

Chapter 6

MARKETING BREAKTHROUGHS

What constitutes a breakthrough?

What role does competitive culture play?

What role does marketing play?

What role does competitive selling play?

How will the next group of industry leading firms break into new markets?

Your Gut is Still Not Smarter Than Your Head: How Disciplined, Fact-Based Marketing Can Drive Extraordinary Growth
Kevin Clancy and Peter Krieg
John Wiley & Sons, 2007

The issue here is growth. The preponderance of management researchers concluded years ago that corporate stasis is impossible to achieve. You are either growing or declining.

If your venture is typical, it is probably expanding in some market segments and atrophying in others, all at the same time. Many of our research partners feel strongly about simply maintaining their position in markets that they've been successful in for years. They invariably ask, "What's wrong with simply maintaining the ground that we've spent years acquiring? It is now predictable, profitable, and we can count on the consistent revenue as the financial basis of our company."

We believe the perception of a stable market segment is an illusion. Every market category we study is in the process of expanding or contracting at all times. And any company that sees their revenue as stable over the years is either grabbing a larger share of a shrinking segment or getting less and less of an expanding one.

It's not good for your soul when the alarm clock goes off in the morning and your first conscious realization is that today you must go to the office, have a cup of coffee, and figure a way to capture a larger share of a shrinking pie. Over the long haul, it's an endlessly frustrating strategy that will eventually prove unsatisfying.

Okay, since we've defined management of the status quo as impossible, it becomes clear that the segments of your company that aren't billowing with new opportunities must be appropriately addressed. They should either be jettisoned and your finite resources invested elsewhere or infused with new energy and focus.

In the AEC world, we live and die by our ability to compete for new work. But all competitions are not the same.

Most firms categorize their new business development activities into these categories:

- **Repeat work:** New work from past clients; no competition required.
- **Extensions:** Current project additions and extensions; no competition required.
- **New work:** New projects for which you must compete.

The ratio of these three contributors to total revenue varies widely depending on the age, size, and specialty of the firm, but if we were forced to manufacture a worldwide average, it would be 25% repeat work, 25% extensions, and 50% new work.

Lateral Marketing: New Techniques for Finding Breakthrough Ideas
Philip Kotler and Fernando Trias de Bes
John Wiley & Sons, 2003

What constitutes a breakthrough?

It goes without saying that current project extensions and repeat work can be extremely profitable endeavors and they cost very little to acquire. Most firms do reasonably well at exploiting these opportunities since they often show up with little additional effort. But this chapter is about how to capture new work. And not just any new work. It's about how to win competitions when you are not the most qualified competitor.

All real growth depends on a firm's capacity to pull off the occasional marketing breakthrough. And even though we are all aware that these wondrous occurrences require an entertaining admixture of art, science, providence, and magic, there are still some rational observations we can make about the process.

For the sake of this discussion, let's use the following definitions:

- **New business development** refers to all accumulated activities involved in acquiring new work.
- **Marketing** includes all actions taken in order to get your face in front of a prospective client.
- **Sales** becomes the issue only once you find yourself in the presence of a prospective buyer.

We have studied many firms which excel at marketing but struggle to win competitions and close sales. Others are masters at competitive selling, but don't get in on their fair share of new leads. In this chapter, we will turn our attention to both endeavors.

What role does competitive culture play?

The Entrepreneurial Mindset: Strategies for Continuously Creating Opportunity in an Age of Uncertainty
Rita Gunther McGrath and Ian C. MacMillan
Harvard Business School Press, 2000

Who, specifically, in your company is responsible for the competitive health of the venture? We've surveyed hundreds of AEC firms and asked that exact question. Typically, a relatively small group of individuals is identified as leading the effort to keep the company competitive. Often it's the principals, the people in senior management, or the marketing group. Unfortunate answer. That's how the last generation did it.

What is a competitive culture?

In every company, employees feel responsible for executing a variety of assignments for which they expect to be evaluated and remunerated. In a culture that values competitiveness, every employee genuinely feels that in concert with their typical job responsibilities, they have accepted an overarching commitment to advance the competitive posture of the company.

In a culture that values competitiveness, every employee genuinely feels that, in concert with their typical job responsibilities, they have accepted an overarching commitment to advance the competitive posture of the company.

A competitive culture won't make you better at sales or closing new work.

What it will do, however, is create a powerful force in the following arenas:

- **Leads:** No matter how hard your marketing department currently works, a group of employees who are highly networked in the community will simply turn up more new project signals.
- **Intelligence:** With the addition of a new web of tentacles into the business community, you will get far more feedback about how your company and your brand is perceived, particularly if you train your people how to ask. You will also hear, with much more accuracy, how your competitors' services are being received.
- **Opportunities:** The business community constantly vibrates with new opportunities and it's tough to take advantage of those you never hear about. Why not employ the eyes and ears of everyone in the firm? Almost anyone can be trained to listen for the sensitive signals that carry the seeds of new business opportunities.

One of the most entertaining and delightful side effects of creating a competitive culture is that new business opportunities and project leads will appear from the most unpredictable and unlikely people. And although some of your employees will be offended and refuse to participate, we have repetitively seen that most will thrive under the new system. Clearly, almost every organization has locked within it a competitive potential that few firms have learned to exploit.

Getting started
Competitive cultures can be inordinately valuable and, once operational, many develop a life of their own. Existing employees will, without prompting, promulgate the system among the new and uninitiated group members. But since we've never seen a competitive culture spontaneously come to life, you are going to have to invest resources to purposefully design and instigate the system.

To get the conversation started, we typically ask every employee: In which markets does your firm compete and what are your competitive advantages in each? The accuracy of the responses is usually depressing, but they will give you an idea of how cognizant your staff is of current market realities.

The next step, obviously, is to teach them. We've seen this executed in a variety of ways, but the following issues seem paramount:

1 Start at the top
All teaching sessions must be led by and attended by people in senior management. Everyone knows that if the new initiative is not important to the leaders of the company, then it's just another management program that eventually won't make any difference.

2 It's everyone's responsibility
It is critical to discuss why the competitive health of the company should be everyone's business. The professionals at most companies have been put through a collection of ill-advised new management regimes and if they sense this is just the latest fad, they will listen politely while looking the other way.

3 Make expectations clear
It is critical to make all expectations perfectly clear. Exactly what do you want your staff to do?

(a) Document their personal business network.
(b) Make it grow.
(c) Use that network to listen for critical market intelligence: (1) new business opportunities, (2) perception of your firm by others, and (3) perception of competitors by others.

All staff must understand that this program is about gathering market intelligence and not singularly taking action. No one is expected to sell, change market perceptions, or close business. Just ask them to listen and report what they hear.

4 Hold regular progress meetings
Each employee must review their activities with the person in senior management assigned to them. These sessions should take place at least four times per year. The more effective programs require that an individual remove any names from their network list that have not been contacted in 60 days. Personal networks are constantly evolving organisms and must be consistently tended. It is common for employees to put individuals on their personal network list that they haven't actually spoken to in years.

5 Assign responsibility to senior management
Specifically assigned people in senior management must be held responsible for not only their own networking, but the business social activities of their assigned staff. Some companies allow principals to compete with each other to determine whose group is producing more actionable market intelligence.

Everyone knows that if a new initiative is not important to the leaders of the company, then it's just another management program that, eventually, won't make any difference.

Marketing Breakthroughs

6 Evaluate performance

How do you intend to keep score? If performance metrics are not developed and made apparent to all, nothing will happen. In fact, nothing will happen very slowly at great expense.

7 Remunerate the achievers

How will achievers in this new system be paid? If additional recognition is not part of the system and no additional pay occurs, then this deal is nothing but talk and the entire staff will treat it accordingly.

In companies that have created a competitive culture, everyone in the venture is cognizant of the business community's landscape of rivalries and is rewarded for the constant mining of new opportunities.

Breakthrough: Stories and Strategies of Radical Innovation
Mark Stefik and Barbara Stefik
MIT Sloan Management Review Press, 2004

What is the culture's role in breakthrough marketing?

It is critical to distinguish between your competitive strategy, the organizational structure you have in place, and the individuals responsible for its execution. It's not unusual to find principals of AEC firms who have authored exciting, relevant competitive strategies that simply have no impact. This phenomenon occurs because most AEC principals are intelligent, perceptive individuals who read about the latest performance trends and understand what they would like to see happen in their company. Interview a large variety of AEC executives and you will find absolutely no shortage of competitive strategies. So why then does a firewall seem to develop between the intentions of senior management and what actually gets executed?

The simple answer is the lack of metrics, the lack of recognition, and the lack of supplemental pay. I know you're tired of hearing me repeat this, but …

you get the performance you measure and the performance you pay for.

Good intentions are heart warming, but they don't powerfully animate human behavior.

The Marketing Gurus: Lessons from the Best Marketing Books of All Time
Chris Murray
Portfolio Hardcover, 2006

What role does marketing play?

It's tough to win new projects that you don't know about. The job of your marketing effort is to make certain that you make it to the presentation phase of every project in which you are interested.

Marketing is the process by which we first learn of new project opportunities and our research tells us that most firms in the AEC industry overwhelmingly utilize the very same sophisticated technique: waiting for the phone to ring. Most successful practitioners spend years creating a personal network of individuals who call the minute they hear of something new. As a matter of fact, in most markets, these rumor networks have become so efficient that all major players will hear of a new building project within 24 hours of the first leak.

Several years ago, I called a corporate vice president to inquire about his new building. He responded by saying, "You people must all live together! The board of directors just approved our new project this morning and I've spent all afternoon answering phone calls from architects, general contractors, interior designers, carpet salesmen, furniture representatives, and one guy who wants to wash the windows when the new building is complete."

> **M**ost people think that marketing is the process by which you get chosen for a job. What we're trying to do in our firm is turn that around 180 degrees and say marketing is not the process by which we get chosen, but the process by which we choose. We try to identify, in advance, strategically, the targets, projects, clients, and opportunities that we would like to be involved with and then we go after those things. We basically reach out and grab those opportunities rather than waiting to be called or approached. Our idea is that if you get an RFP in the mail then it's too late. If you hear about it word of mouth, it's too late. You really have to be proactive and you have to be choosing your clients rather than waiting to be chosen.
>
> SCOTT SIMPSON, FAIA, SENIOR PRINCIPAL FOR STRATEGIC GROWTH
> KLINGSTUBBINS
> BOSTON, MASSACHUSETTS

The object of the exercise is not to sharpen your early warning system, but to be in the room when a client makes the decision to build. By the time the hearsay experts swing into action, the client's world becomes incredibly noisy and their selection process often responds by becoming convoluted and irrational.

So, what's required to get on the short list?

The win rate of the world's greatest sales team is somewhat hampered if they never make it into the presentation room. Since the object of all marketing activities is to get on the short list, we have spent time asking AEC firms of all sizes what seems to work for them.

Making it to the final show, once again, is a complicated assemblage of skill, persistence, and good fortune. The most powerful impactors seem to be:

1. Previous relationships
2. Centers of influence
3. Brand

All three are well-known concepts and don't warrant much discussion here, except to say that the identification and networking of people who don't actually make hiring decisions, but actively influence those who do, is an important endeavor, the power of which can't be overstated.

What role does competitive selling play?

The discipline of competitive selling and its role in breakthrough marketing is truly a curious phenomenon. It's very rare to find a company in the AEC industry that seems to be doing it optimally. We see many firms that should spend less time on selling techniques and more time fabricating a competitive company. On the other hand, it is common to find experienced, credentialed, industry-leading companies that regularly lose projects to lesser competitors because they simply refuse to invest adequate energy in aggressive, effective, competitive selling.

Many times we begin a discussion on competitive selling by asking the sales team: "Of all the information that you must communicate during a sales presentation, what do you consider the most important?" Usually they say things like: "We want to explain who we are as a company, what our values are, what our credentials are, and why we are the best choice for this project."

Wrong answer.

There are obviously many interlaced strategies that must coalesce in order to achieve above average success at winning competitions, but the march to dominant selling begins with this specific attitude:

It's not about you.

Identification and networking of people who don't actually make hiring decisions, but actively influence those who do, is an important endeavor, the power of which can't be overstated.

Marketing Breakthroughs

> In many competitive situations, we will discover an issue that suddenly becomes a need and a hot topic of conversation. By digging deeper and asking more questions before the presentation phase, we can often learn of subtleties that our competitors simply don't know about. We also concentrate on describing needs that our prospective clients may not know they have.
>
> MARGARET BOWKER, ASSISTANT VICE PRESIDENT
> JE DUNN CONSTRUCTION
> KANSAS CITY, MISSOURI

While you are standing at the head of the darkened room, proudly displaying the visuals that demonstrate your creative and technical superiority, your prospective client, with eyes progressively glazing, is wondering about two things: (1) Does this presenter actually understand what I'm trying to get done? and (2) If I hire his firm, will I still have my job when this project is complete?

No matter how utterly magnificent your firm is, the potential client you're talking to is worried about his own issues, not yours.

> As listeners, we hear what our prospective clients want to achieve through their building projects. Once we've heard our prospects, we engage them with stories that demonstrate our know-how in related applications. In competitive selling, a story is much more telling of experience and capability than a list of achievements.
>
> HOWARD TELLEPSEN, CHAIRMAN
> TELLEPSEN BUILDERS
> HOUSTON, TEXAS

While you are standing at the head of the darkened room, proudly displaying the visuals that demonstrate your creative and technical superiority, your prospective client, with eyes progressively glazing, is wondering about two things: (1) Does this presenter actually understand what I'm trying to get done? and (2) If I hire his firm, will I still have my job when this project is complete?

What do potential clients respond to?
During the typical give and take of the professional selection process, clients are usually confronted with a daunting variety of approaches and messages. Obviously, some of these messages penetrate better than others. So, which ones seem to carry more punch? Here's a list of eight topics worthy of attention:

1 Company brand
What are you known for? What impressions have preceded you? Remember, name recognition is of no value if it is not instantly associated with positive characteristics.

2 Credentials with the specific type of project being considered
You may have an enviable brand, but what do you know about the type of project under consideration?

> At the end of the day, we have found that we don't win projects based on the length of our portfolio. We win them based on chemistry and whether the client feels the trust that we'll take care of them. When we lose projects, that's why we lose them. Chemistry. Rarely do we lose because we didn't have the right answer. It's usually because of relationships or chemistry.
>
> GREG NOOK, EXECUTIVE VICE PRESIDENT
> JE DUNN CONSTRUCTION
> KANSAS CITY, MISSOURI

3 Apparent understanding of the project being considered
Besides your credentials with this particular type of project, has your team specifically addressed the issues at hand? Have you learned anything about the prospective client's prejudices that are not common knowledge? If you base your approach on the same assumptions as the other competitors, differentiation will be difficult to display.

4 Innovative approaches
Have you got anything to say about the proposed project that distinguishes you from the other competitors? Innovative approaches that solve problems considered by the client to be

of low value won't carry the day. We've witnessed way too many examples of credentialed professionals who fall in love with their own approach and repetitively attempt to apply it in inappropriate circumstances.

> I think that when you're trying to either enter a new market or dominate a market, there is so little research done in the AEC industry that any firm that puts their mind to how to perform better will have a distinct advantage. The history of CRS is an example of that. Bill Caudill wrote his thesis at MIT on how to design better schools and before they'd ever designed a school he published this document on the impact of school facilities on education. The next thing you know, CRS was the national expert on school design.
>
> CHARLES B. THOMSEN
> FORMER CHAIRMAN, 3D/I
> & ADVISORY DIRECTOR, PARSONS
> HOUSTON, TEXAS

5 Comprehensive response to the entire proposal process
Most competitive new business development processes require quite a variety of interactions between prospective clients and the professionals they are considering hiring. Each single contact, phone call, letter, email, brochure, meeting, lunch, and presentation all add to the overall impression. The most effective competitors see this process as a single comprehensive entity that must be holistically managed.

6 Responsiveness to client's emotional drivers
As a youngster, I was on a team asked to execute a comprehensive facility analysis for a large integrated energy company in Houston. We worked for three months dissecting all corporate functions and helping our client understand the relationship between their individual business units and the facilities that might best house them. Modestly, I thought our report was staggeringly impressive. It contained a "sociogram," which at the time was leading edge. We mapped the information that flowed into and out of every employee in this gigantic company and displayed it on the

boardroom wall. For many of the directors, it must have been the first time they ever saw how the humans in their company actually process their work.

The entire board of directors, sitting around a 24-foot marble table, listened to our presentation very politely. We explained that their company, in order to accommodate projected growth, needed to build 1.2 million square feet of new buildings which could best be accomplished with an integrated low-rise campus plan.

At this point, our team was beaming with self-pride. We had just analyzed a complex situation and described a scenario that would serve this client handsomely.

A dead silence followed while the chairman, a physically and verbally imposing industry giant, deliberately leaned forward and hunkered down over his microphone. He ponderously announced: "Gentlemen, we're going to build the tallest damn building in this part of the country. Meeting adjourned."

So there you go. All of that analysis and we forgot to delve deeply enough to uncover the powerful emotional drivers that would carry the day. The chairman's need to pile steel and glass higher than anyone else had nothing to do with effectively housing his employees. His motives were entirely emotional, but he would never admit to such a conclusion.

> Here's the most important thing: You have to portray to the client the absolute commitment that you're going to take care of his needs, that his project is going to get your full attention, your full focus, and that you have the team in place to get it done. Attitude, enthusiasm, and commitment are incredibly important sales attributes that can sometimes trump experience, sometimes trump design ability, and sometimes even trump fee levels.
>
> SCOTT SIMPSON, SENIOR PRINCIPAL
> KLINGSTUBBINS
> BOSTON, MASSACHUSETTS

People's motives are entirely emotional. The obvious problem is that even though emotional motivators are clearly paramount, most people won't honestly express them. Therefore you must learn to read them implicitly.

The obvious problem is that even though emotional motivators are clearly paramount, most people won't honestly express them. Therefore you must learn to read them implicitly.

7 Likeability of the sales team

As disappointing as it sounds, humans still have a strong tendency to buy from people they like. We have all experienced professionals who have little else to offer other than likeability and, thank goodness, they seldom win on friendliness alone, but it's a mistake to underestimate the power of high communicators. A compelling message delivered by an emotionally committed speaker can sometimes work wonders … and cause calamities.

8 Fees and pricing

Nobody likes to compete on fees. Period. We all have learned that if it comes down to price, we've failed miserably to differentiate

Nobody likes to compete on fees. Period. We all have learned that if it comes down to price, we've failed miserably to differentiate ourselves. And worse, we've failed to establish, in the mind of the buyer, the value of our work.

ourselves. And worse, we've failed to establish, in the mind of the buyer, the value of our work.

Typical competitive selling errors
Repetitive buyers of AEC products and services describe the following as the most commonly repeated mistakes made during the competitive selling process:

1. Thinking that your credentials will carry the day.
2. Assuming that your personal relationships will carry the day.
3. Expecting that your experienced sales team will win.
4. Assuming that what has worked in the past will work again.
5. Ignoring the emotional drivers of your potential buyer.

How will the next group of industry leading firms break into new markets?

If we set aside all of the philosophy and get right to the point, it becomes painfully clear that the single most compelling change in the development of new business for the next generation of market leaders will be that all of their processes will evolve from the intuitive to the purposeful. Specifically, they will:

How Breakthroughs Happen: The Surprising Truth About How Companies Innovate
Andrew Hargadon
Harvard Business School Press, 2003

1 Create competitive cultures
Creating a competitive culture will require substantial effort but the benefits will massively outweigh the costs. The next generation of market leading firms will create a competitive culture that permeates every functional element of the company. All employees will see the competitive health of the venture to be their personal charge.

They will be taught how to gather critical market intelligence and publicly rewarded for doing so.

2 Generate their own leads
There is nothing wrong with having your personal contacts call with the latest new project lead. The next generation will supplement this passive approach, however, with an aggressive program to take their message directly to the targeted audience. These institutionalized marketing programs will not only provide more

accurate intelligence, they will put the practitioners in a much stronger networking position.

3 Create a competitive selling team
Competitive selling is still more art than science, but nevertheless, it should not be treated as a part-time hobby by those responsible for generating new business.

4 Undertake rigorous pre-competition analysis
Many experts feel that the game is usually won by the group that most perceptively prepares, knows more about details of the assignment, understands more about exactly what the potential client is trying to achieve. We have repetitively observed that as AEC companies grow, they tend to become more and more impressed with themselves and their credentials. These groups are particularly vulnerable to smaller, more energetic competitors who are simply willing to work harder to close an important project.

> When we were pursuing a contract with the State of Mississippi for their intelligent transportation services, we put together a way of communicating with them over a period of time to where by the time we got to the presentations, we were actually into solving their problem as opposed to presenting our qualifications.
>
> JAMES BEARDEN, CEO
> GRESHAM, SMITH AND PARTNERS
> NASHVILLE, TENNESSEE

5 Exploit professional joint ventures
All major players have learned by now that the quickest and most efficient way to beef up your credentials is to joint venture with a complementary firm. Nothing about this strategy is new and it is often successful. In reviewing with our research partners a series of joint venture arrangements, it seems that practically all problems could have been avoided by properly negotiating clear agreements and anticipating the standard human management problems which seem to plague most JV endeavors.

6 Deal effectively with emotional drivers

In younger years, I used to believe that humans were rational beings with the occasional emotional experience. That is clearly not the case. People are distinctly emotional beings who have the occasional rational experience. The problem with attempting to ferret out the emotional motives of your prospective clients is that they will seldom tell you the truth. Therefore, we have no choice but to implicitly infer what is really behind their decisions.

7 Create new integrated project alliances

Here we see a robust collection of opportunities for every company in the AEC industry to either stumble or advance. Everyone knows that current project delivery systems are broken and most substantial buyers around the world are actively investigating alternative strategies.

This particular topic has been the subject of research at the Rice University Building Institute for the past several years and, while the exact future of project delivery is somewhat uncertain, there is one thing we know for sure: Tomorrow's clients will demand interdisciplinary teams of professionals willing to integrate their expertise, share risk, and devise processes that are less time-consuming and more responsive to the increasing vagaries of world commerce. Companies who take the safe route and wait to see how these new alliances develop may find themselves behind the power curve. It's time to innovate.

8 Align facility and business strategies

AEC professionals who continue to think that their industry is only, simply, about designing and building buildings are missing the real opportunities. The next generation of architects will see their mission as solving business and societal problems utilizing their knowledge of the built environment. The future will belong to those who can sit down with the CEO of a major corporation, have an intelligent conversation about the full range of their business strategies, and then explain how the buildings can positively impact business outcomes. The same phenomenon applies to those interested in having positive societal impacts through planning, designing, and building not just buildings but whole communities as well.

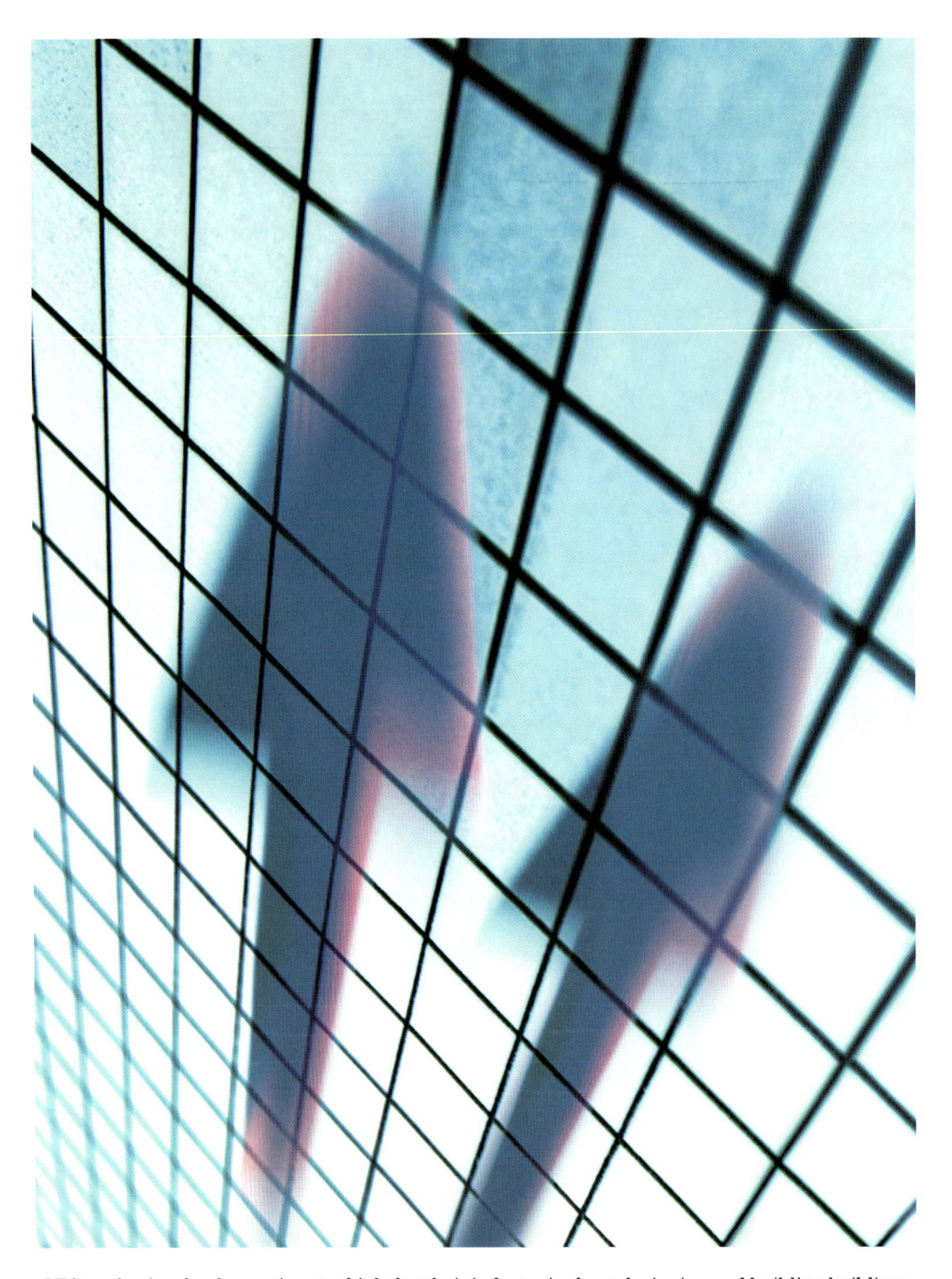

AEC professionals who continue to think that their industry is about designing and building buildings are missing the point and will sooner or later pay the price. The next generation will see their mission as solving business and cultural problems, utilizing their knowledge of the built environment.

> Generally speaking, I think the building industry does a reasonably good job of researching the performance of buildings' systems, such as energy efficiency, where we're building a body of knowledge. But what's lacking is a documented, researched understanding of how a building or facility can enhance the performance of a business enterprise.
>
> ---
>
> RK STEWART, FAIA, HON FRAIC
> 2007 PRESIDENT, AMERICAN INSTITUTE OF ARCHITECTS
> & PRINCIPAL, GENSLER
> SAN FRANCISCO, CALIFORNIA

Effectively and creatively shaping the built environment has never been more important than it is today, but tomorrow's victories will belong to those capable of looking beneath the obvious and understanding the real value of our work.

High-Impact People

Less than acceptable competitive results, which are rampant, are typically not caused by ineffective strategies. They are caused by people and processes that simply aren't up to the task. And since your people create your processes, it's clear to us where to look when it's time to improve performance.

Chapter 7

HIGH-IMPACT PEOPLE

Who are high-impact people?

How do you effectively integrate them into your work process?

How do you keep them?

> If each of us hires people who are smaller than we are, we shall become a company of dwarfs. But if each of us hires people who are bigger than we are, we shall become a company of giants.
>
> DAVID OGILVY (1911–1999), FOUNDER
> OGILVY & MATHER

The AEC world is one that can be characterized by the ever-present, unrelenting search for competitive advantage. So, it's no wonder that architects, engineers, and builders have so much energy and intellectual capital invested in the development of competitive strategies.

Here's an interesting observation: We have yet to find a strategy that isn't rational and executable. Some are even perceptive. Here's another interesting observation: They are remarkably similar.

Even though we have observed firm leaders who possess a wide range of personalities and have witnessed them dealing with divergent external conditions, we see absolutely no shortage of perfectly acceptable competitive approaches. This being the case, why do so many of these well thought-out and well-funded programs fail to meet their intended outcomes?

In a word, execution. Less than acceptable competitive results, which are rampant, are typically not caused by ineffective strategies. They are caused by people and processes that simply aren't up to the task. And since your people create your processes, it's clear to us where to look when it's time to improve performance.

Firms that are serious about their competitive posture sooner or later do some sort of analysis of the firms with which they compete. Even though we find the quality of these analyses somewhat spotty, the concept is right on target. Where AEC firms typically underperform other industries is how well they evaluate the performance of their own professionals.

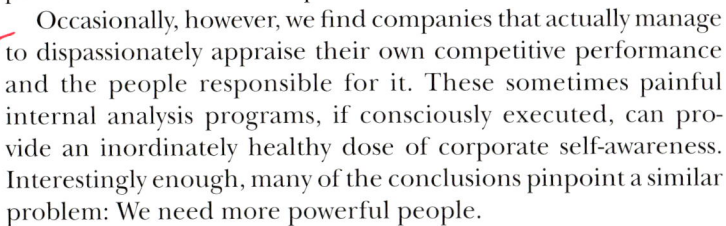

Occasionally, however, we find companies that actually manage to dispassionately appraise their own competitive performance and the people responsible for it. These sometimes painful internal analysis programs, if consciously executed, can provide an inordinately healthy dose of corporate self-awareness. Interestingly enough, many of the conclusions pinpoint a similar problem: We need more powerful people.

All AEC companies report that they are, in fact, in the knowledge business and that, while process is critical, the professional expertise they sell is living between the ears of their people.

Actually, we believe that the AEC industry isn't strictly in the knowledge business, we're in the knowledge delivery business. And what is the only mechanism at our disposal for delivering our knowledge to our clients? People. And which people shoulder the net professional value of the venture?

Execution: The Discipline of Getting Things Done
Bossidy, Charan, Burck
Crown Business, 2002

People who outperform the competition.

> **I**n today's energetic market, the biggest area of competition is not for projects, but for talented and committed staff.
>
> FREDERICK FIRLOTTE, PRESIDENT & CEO
> GOLDER ASSOCIATES
> MONTREAL, CANADA

Who are high-impact people?

There is an infinite variety of definitions available to describe the term "high-impact people" and there aren't many topics on this planet that have been more studied and documented and published. It's interesting that with so much analysis by so many smart people, the topic of why humans do what they do is still, in many ways, a puzzlement.

Talent Management Systems: Best Practices in Technology Solutions for Workforce Planning
Allen Schweyer
John Wiley & Sons, 2004

> High-impact people are not necessarily the innovators or motivators, although those are important. For us, high-impact people are those who inspire confidence both inside the company and outside. It's only through these people that clients keep coming back. Because Gleeds is a service company, a lot of clients come back because they trust Gleeds, and a high-impact person is someone who can create that trust.
>
> ---
>
> DAVID KELLY, MANAGING DIRECTOR
> GLEEDS AUSTRALIA
> SYDNEY, AUSTRALIA

When I was younger, I was fortunate enough to witness a European multinational conglomerate decide that it wanted to own a large American commercial construction company. After searching for one to buy, the Europeans made the decision to fabricate a new venture from scratch. The conglomerate then assembled a small group of leading industrial psychologists, organizational behaviorists, and management consultants and told them to acquire the best construction talent available. Money was plentiful. The conglomerate committed to invest $200 million per year for five years. After that, the new company would be expected to function on its own, profitably.

It was amazing to see that much capital and that much expertise focused on creating a new venture from scratch. And it worked. Mostly.

The development strategy was clear. Start by acquiring the best possible talent, who, in turn, would create the processes and procedures required to breathe life into this new organization.

The experts set about to pull together a senior management team and, in doing so, applied all of the latest selection and evaluation strategies available. As a serial entrepreneur accustomed to building new ventures by the seat of my pants, I was thrilled to witness how all of these experts with all of this money went about their quest. I was also amused to learn that one of the initial lead engineers hired, during his first week of employment, was arrested by the Houston Police Department and jailed for exposing himself at the local mall. So there you go. All of that money and all of those experts and they still managed to recruit a felon.

The Europeans, however, were clever, diligent, and invested wisely. They clearly knew how to initiate a new venture in a different culture. It went so well, in fact, that the conglomerate only

While current knowledge of human behavior is genuinely interesting and there is certainly an impressive collection of intelligent experts working the problem, the task of acquiring and nurturing high-impact professionals remains complicated, confusing, disorienting, and fascinating.

held the start-up for four years before selling it, for a hefty profit to the management group that they had hired in the first place.

Here's what I took away from watching this complex tangle of decisions applied to company building: While current knowledge of human behavior is genuinely interesting and there are certainly an impressive collection of intelligent experts working the problem, the task of acquiring and nurturing high-impact professionals remains complicated, confusing, disorienting, and fascinating.

Harvard Business Review: On Finding and Keeping the Best People
Peter Capelli and Ibarra Hermina
Harvard Business School Press, 2001

> **W**e believe that clients hire people, not companies. We cultivate our brand, not so much because it attracts clients, but because it attracts the best people.
>
> RICHARD MORRIS, EXECUTIVE VICE PRESIDENT
> SMITH SECKMAN REID, INC.
> HOUSTON, TEXAS

So, if we combine lessons from the general literature with lessons from hundreds of interviews, we can offer the following completely non-scientific observations about high-impact people in the AEC industry.

1 They are emotionally intelligent

This term has taken on a life of its own over the last couple of years. For our purposes, emotional intelligence describes an individual's ability to understand the feeling states of other people. And since humans are driven far more by their emotions than their intellects, the capacity to understand emotional issues lies at the very heart of effective leadership and teamwork. Some people are gifted at identifying how others are feeling, and some people don't have a clue. There are people in your office who can recognize and relate to the feelings of their co-workers, while others are so completely consumed with their own emotional baggage that they are only minimally concerned with the feelings of others. Since humans relate to each other, even at the office, far more emotionally than intellectually, powerful leaders invariably operate with elevated high levels of emotional intelligence.

> **T**he high-impact people we're looking for are not those that can grace a room and befriend everyone there, not that that's not important, but we really look at the high achievers as the ones that can materially impact the performance of our project teams. For the most part, our most valuable people recognize they can't achieve on their own, that the only way they can excel is with a good team, with a good group of people around them.
>
> AL STOWKOWY, PRESIDENT
> STUART OLSON CONSTRUCTION
> CALGARY, CANADA

2 They are self-aware

If "emotional intelligence" is about understanding and relating to the feelings of others, high achievers must have access to the same flow of consciousness about themselves. High-performance people are in the business of positively influencing the attitudes and behavior of their team members, an undertaking that is virtually impossible without an adequate dose of self-awareness. We have all witnessed what happens when an intelligent, motivated team member substantially misinterprets their own capabilities. They usually produce more human discord than group advancement. And while, today, there is an interesting collection of psychological programs available designed to train people on matters of teamwork, group interaction, and self-awareness, you must admit, it's simply more efficient to select people who show up with the ability to emotionally relate to others.

Powerful Conversations: How High Impact People Communicate
Phil Harkins
McGraw-Hill, 1999

3 They are self-motivated

External motivation for high-impact people is seldom necessary. They bring their own. Management's task, therefore, is not to provide more motivation, but to understand the individual actuators that are already in place. Some strong employees arrive with motivators that are clear, positive, and reasonable, a condition that should be heartily reinforced. Others will be driven by sources that are less in concert with group culture. These people, if not quickly redirected, can do harm to team culture and functioning. There is, of course, a substantial body of literature reporting that efforts to redirect an individual's motives can usually be expected to fail. The more efficient approach is to learn what drives each of your high achievers, keep the ones that already match your culture and eliminate the others.

> Many high-impact people are not very detail oriented. They have great ideas, a lot of passion, but you have to be sure they're teamed up with support mechanisms to help their ideas come to fruition. The worst thing you can do is let them run around in circles with all of that energy.
>
> RICHARD MORRIS, EXECUTIVE VICE PRESIDENT
> SMITH SECKMAN REID, INC.
> HOUSTON, TEXAS

External motivation for high-impact people is seldom necessary. They bring their own.

4 They have superior knowledge and can deliver it

The knowledge, experience, and judgment required to design and build a complex project can feel somewhat overwhelming. It's hard to image defining any professional as high impact unless they bring an unusual level of expertise to the game. But as discussed earlier,

we aren't in the knowledge business. We're in the knowledge delivery business.

We've all worked with brilliant people whose interpersonal communication skills are so ragged as to render them useless. On the other hand, we have all worked with gifted communicators who, in the final analysis, actually have nothing to say. What a lousy set of options. (A) Knowledgeable people who can't effectively deliver their expertise or (B) empty suits who suck the air out of every room by demanding to be the center of attention. In walks the high-impact player. They actually possess powerful professional skills and have the interpersonal relationship capabilities to rally others to their quest.

The Fifth Discipline: The Art and Practice of the Learning Organization
Peter M. Senge
Currency Doubleday, 2006

5 They seek responsibility

In any team sport, sooner or later, competitive events conspire to create a crisis moment—an instant in which the game will either be won or lost. When that pressure-packed moment arrives, most team members know who among them will perform best. In such times, high-impact people want to be in charge.

The problem for management, of course, is knowing how much responsibility to dole out and when. Too much too soon creates fertile conditions for failure, while too little will frustrate talented people.

6 They need rapid feedback

Everyone on the planet seems to be suffering from attention deficit disorder these days. Nevertheless, we all know it's a good idea to provide each employee with regular performance evaluation. With high-impact players, consistent, rapid feedback is even more critical. Obviously, clear operational metrics must be agreed to, but don't leave a bunch of hard-chargers to their own devices for long. The fact that they move quickly means that they have the capacity to turn a small error in judgment into a perfect disaster in no time at all.

In addition, many high achievers seem to be installed with an elevated capacity to self-criticize, which management must monitor and attempt to mitigate.

How do you effectively integrate them into your work process?

In his book, *Good to Great*, Jim Collins discusses the interesting relationship between company strategy and the high-impact employees responsible for its execution. Traditional wisdom tells us that senior management should first decide the direction in which they want to take the company and then set about to find the right people who can make it happen. Collins advances the view that the more effective approach is to go get the best people

Good to Great: Why Some Companies Make the Leap and Others Don't
Jim Collins
Harper Collins, 2001

you can find and let them help define the firm's direction. In other words,

it's not about the strategy, it's about the people ... get them first.

> We believe our company success to be about 80% due to people and about 20% due to process; you improve the people, you improve the company.
>
> DOUG HARRISON, MANAGER OF ADMINISTRATION
> STUART OLSON CONSTRUCTION
> CALGARY, CANADA

The integration of talented people into existing teams and work processes must be accomplished with great care. Properly done, these people can produce amazing results. Improperly integrated, they will quickly and efficiently tear up previously productive teams.

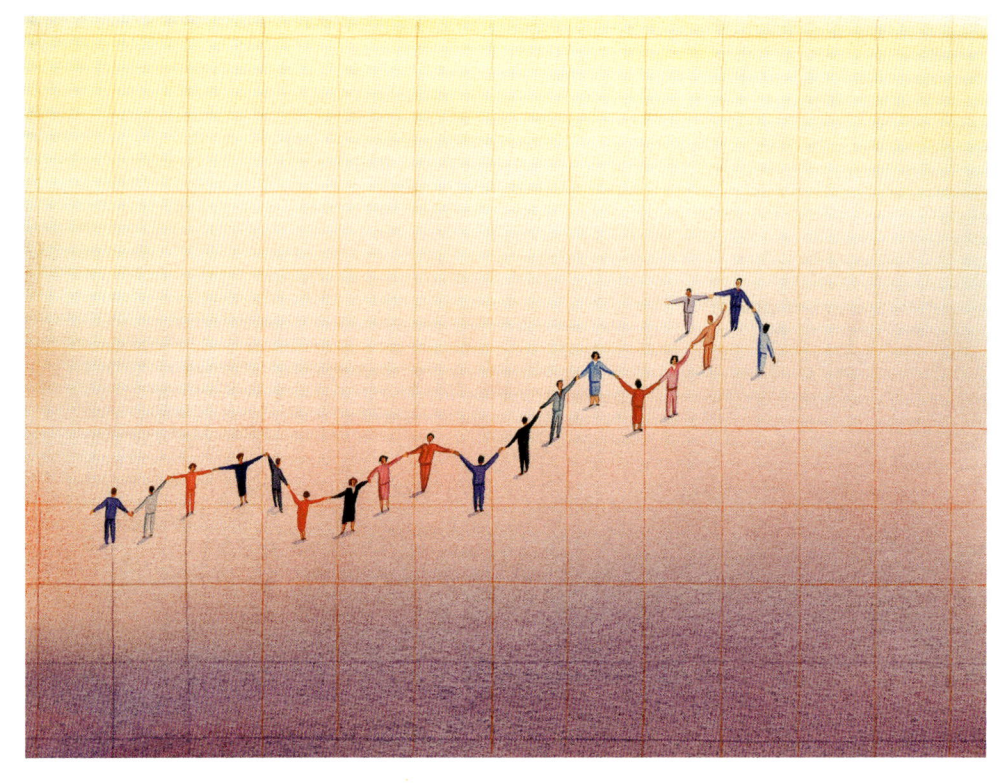

The integration of talented people into existing teams and work processes must be accomplished with great care. Properly done, these people can produce amazing results. Improperly done, they will quickly and efficiently tear up previously productive teams.

Our interviews have provided the following observations:

1 Identify them

The superior performance of some high achievers is readily visible to all, while others seem to do what they do with more subtlety. Step one in the process of successfully exploiting the talents of special people is to accurately identify them. At Turner Construction, 110 years old with 44 offices in 12 countries, senior management, as part of their "development appraisal system," annually creates a list of "high potential" employees. "Learning maps" are developed for each of them and they are purposefully assigned to a wide variety of divergent projects. The object is to create the next generation of corporate leaders.

Companies that haven't developed a comprehensive process for identifying and communicating with their high achievers are subject to the surprise defection. From time to time you will lose good people, but if it happens unexpectedly, management has suffered a lapse.

> Our high potential people are all high maintenance. They typically want more responsibility and demand more attention from management. Retaining these people is obviously high priority for us and we've learned that they respond to growth opportunities even more than pay.
>
> ROBERT D. LEVINE, SENIOR VICE PRESIDENT
> TURNER CONSTRUCTION
> NEW YORK, NEW YORK

2 Position them

As we discussed earlier, external motivation is typically not required. The trick seems to be the constant renegotiation of the relationship between the high achiever's needs for autonomy with the group's need for effective collaboration. Left to their own devices, many talented people will invest way too much time attending to their own professional needs while paying short shrift to those of the group. On the other hand, if management is successful at getting its special people to totally acquiesce to team needs, you will probably never enjoy the benefits of their unusual capabilities. Obviously, an effective compromise is called for, but that special equation, if ever found, won't last very long. It is a never-ending endeavor, but one that, when applied to genuine high-impact people, should be well worth the effort.

> At Cannon, we have a high-impact design principal in our Los Angeles office, Mehrdad Yazdani. Even though he is quite an individualist and is widely recognized for that, we need for him to function within the bounds of our large corporate architectural practice. Our solution was to carve out a group called the Yazdani Studio. It allows Mehrdad to be individually recognized for his work and to enjoy a level of autonomy, while allowing us to provide our corporate clients with leading-edge design.
>
> LEE BRENNAN, PRINCIPAL
>
> CANNON DESIGN
>
> LOS ANGELES, CALIFORNIA

3 Focus them

This is perhaps management's most challenging task when interacting with high-impact people. When we asked unusually talented people what they want from senior management, their responses coalesce very tightly around one topic: a clear path for advancement. And by advancement, they mean more responsibility, more authority, more recognition, more money, and eventually ownership in the firm.

Senior management people at almost all firms agree that there should be a variety of paths to ownership and, when we review those paths, all seem perfectly reasonable. The critical issue here is not how the path to advancement is designed. It is whether or not the way has been make perfectly clear to your selected group of high-potential people. This condition is extremely easy to evaluate. First, ask corporate leadership who their high achievers are. Next, interview each selected employee and ask them their understanding of exactly what is required to become an owner or partner in the venture. What we typically find is a confusing hodge-podge of responses.

You can't expect your best people to effectively navigate the course to the top if you don't make the way-finding exceedingly clear.

> We cultivate two kinds of high-impact people: First, the rainmakers who keep bringing in the work and stay involved with the project just long enough that the client doesn't think they're dealing with a salesman, and second, the producers, first-rate engineers who can deliver world-class solutions on time. Either one of these high achievers isn't much good without each other.
>
> WILLIAM CARETSKY, SENIOR VICE PRESIDENT
> SYSKA & HENNESSEY ENGINEERS
> NEW YORK, NEW YORK

The Leader as Motivator: Inspire Great Performance from Yourself and Your Team
Thomas A. Stewart et al.
Harvard Business School Press, 2007

4 Recognize them

What motivates humans to achieve? Scientists, philosophers, and charlatans have been trying to figure this out for centuries. Money obviously works, but public recognition is less expensive and often creates better results.

The object is to create a recognition program that has actual meaning. Companies that get the least impact from these programs are the ones that do it only sporadically and skimp on the ceremonial aspect.

Recognition can occur in many forms, so we asked a group of selected high achievers from all AEC disciplines the following question: Assuming you are being paid fairly, which non-financial motivators do you consider the most valuable? Here are their responses in order of importance:

(a) Personal attention from company leaders to develop a career learning and advancement plan.
(b) Assignment to prestigious teams and projects.
(c) Public recognition ceremonies.
(d) Upgraded workspace.
(e) Benefits that can be shared with family and friends, such as club memberships and travel opportunities.

> Promotion, freedom, and acknowledgment of their work are the sorts of things we offer our identified high achievers, and they are much more effective than money.
>
> GREGORY HODKINSON, CHAIRMAN
> ARUP AMERICAS
> NEW YORK, NEW YORK

Evaluate Them on Teamwork

The object of the exercise here is not to build a loose confederacy of hard-chargers, but to formulate an organization functioning with overlapping high-performance teams. We have found that the most dynamic companies mitigate the powerful individual's natural propensity for singular advancement by making at least a portion of their performance review based on the achievement of the teams to which they are assigned.

5 Evaluate them on teamwork

The object of the exercise here is not to build a loose confederacy of hard-chargers, but to formulate an organization functioning with overlapping high-performance teams. We have found that the most dynamic companies mitigate the powerful individual's natural propensity for singular advancement by making at least a portion of their performance review based on the achievement of the teams to which they are assigned. When dealing with hard-chargers, it is imperative that they be made to understand that their ultimate professional wellbeing won't be based on what they accomplish individually, but what they can produce through collaborating with others.

5 Beware of misdiagnoses

During the past three years, we have gathered a colorful collection of episodes brought on by the temporary misdiagnosis of high achievers. Evidently, it is not unusual for individuals to show up who embody all of the outward characteristics of people who can make things happen. They effectively and creatively utilize all of the latest jargon. They dress powerfully. They are on a first-name basis with other important people. They are members of prestigious social clubs. They play better than average tennis and golf. And they masterfully ingratiate themselves with all of the proper corporate leaders. Here's what happens when they are given a major assignment: Nothing.

But they accomplish 'nothing' with such aplomb and grace that it takes several years for some in senior management to appreciate the full depth of their professional impotence. We've witnessed this occurrence many times, particularly in smaller companies and those ventures that haven't yet developed specific performance metrics against which to measure an individual's effectiveness.

Finding, Hiring, and Keeping Peak Performers
Harry Chambers
Basic Books, 2001

The longer firms let these loveable distractions hang around, the more traumatic and expensive it becomes to part with them.

> We try to hire outstanding people whether or not we have a specific spot for them at the time they become available. A few years ago, we were remodeling a university chemistry building and the client assigned a PhD chemist to represent the building user. He had little, if any, experience with design and construction, but he and we were mutually impressed with how we each performed on the project. Soon after, he joined Linbeck. He has swiftly become an outstanding manager.
>
> MELVIN HILDEBRANDT, FORMER PRESIDENT
> LINBECK
> HOUSTON, TEXAS

How do you keep them?

Most companies and most industries spend substantial energy managing the acquisition of those commodities that they perceive to be in the shortest supply and, therefore, the most valuable. A brief glance at the world's business literature, with its headlines about the performance of major stock markets, private equity acquisitions, hedge fund behavior, and CEO stock options would imply that the commodity in our system that is the most valuable is capital. Every new MBA who is paying attention quickly deduces that the fastest path to obscene wealth is not through designing something or building something. It's through working for a company that moves capital from one place to another while grabbing a small piece to keep.

We believe that sooner or later, many companies will wake up to the fact that

keeping investors happy is not as important as keeping the high achievers happy whose work, in fact, keeps investors happy.

Keeping the People Who Keep You in Business: 24 Ways to Hang on to Your Most Valuable Talent
F. Leigh Branham
AMACOM, 2001

> We want to have space for the people who will make huge differences and we have, therefore, developed various systems to recognize and nurture them. In one such effort, called the Fellowship Program, we identify those who perform at the very top of our profession. Along with the title of Fellow, they get a certain latitude to work broadly on projects throughout the company.
>
> GREGORY HODKINSON, CHAIRMAN
> ARUP AMERICAS
> NEW YORK, NEW YORK

Keeping Your Valuable Employees: Retention Strategies for Your Most Important Resource
Suzanne Dibble
John Wiley & Sons, 2001

So, other than pay, what does it take to keep the people who make the greatest impact? Not a simple issue, and it clearly it depends on the individual, but here's the current thinking of our research partners:

Personal engagement
It is critical that high achievers feel personally understood by senior management and socially involved in the company.

Emotional engagement
High achievers would rather work for a cause than a company. Engage them in an emotional quest for something more meaningful than typical work.

Career engagement
Talented people all want to advance. Many AEC organizations prioritize individual professional learning and career planning. Of particular importance is a program which marks a path to ownership. The companies that don't actively facilitate personal growth often find that high achievers proceed without them.

> It's about ownership, inclusion, a sense of influence, a sense of career pathing. It's not just about promotions, it's about painting a picture of their life at SSR. The unknown, not knowing how you're included, not knowing where you're likely to be in ten years, can create anxiety and that's one reason people leave. They don't have a clear sense of their future.
>
> RICHARD MORRIS, EXECUTIVE VICE PRESIDENT
> SMITH SECKMAN REID, INC.
> HOUSTON, TEXAS

Support
Highly motivated people are all pursuing intriguing agendas designed for personal and professional development. Substantial commitment is created by employers when they are seen by the individual as facilitating their growth. Most high-impact workers we interviewed felt that their managers have a good sense of what the company needs from them, but less of a feel concerning what the individual needs from the company.

Status
Most people don't like discussing it, since it seems a bit self-serving, but the acquisition of personal status, in any of its forms, is powerful medicine for keeping talented people.

Recognition
We've already discussed public recognition. It's powerful and it works.

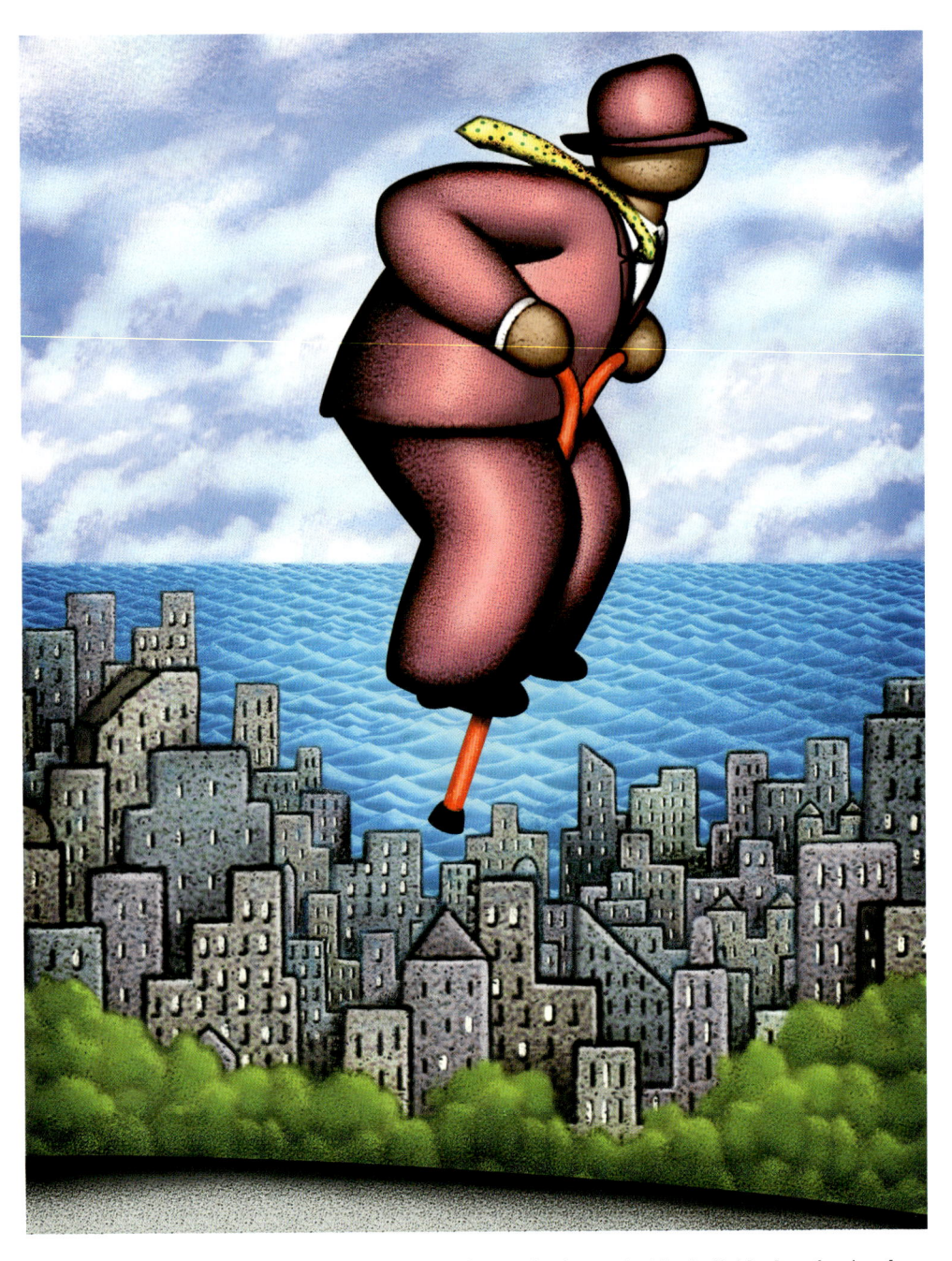

Talented people all want to advance. Many AEC organizations prioritize individual professional learning and career planning. Of particular importance is a program which marks a path to ownership. The companies that don't actively facilitate personal growth often find that high achievers proceed without them.

Flexibility
High achievers typically have a broad array of professional interests, and while they are perfectly willing to focus on the assignments given to them, most respond very positively to the opportunity to gain exposure to divergent skill sets. Companies that actively cross-train report enviable results in retaining their best people.

All companies lose good people so we asked our research partners how they would describe the most common causes. Many groups utilize exit interviews to gather this data and a few of the larger corporations, pursuing more insightful information, hire outside consultants. The vast majority of responses grouped around five issues:

1 They sense minimal opportunities for advancement.
2 They feel under-utilized and under-recognized.
3 They feel the value of their contribution is greater than indicated by their pay.
4 They feel socially or professionally estranged from the group.
5 They simply found a better opportunity elsewhere.

High-impact people and high-performance teams live at the very heart of any company's capacity to thrive in a competitive environment. Many AEC firms have instigated remarkably effective programs to nurture their most valuable asset: Their people.

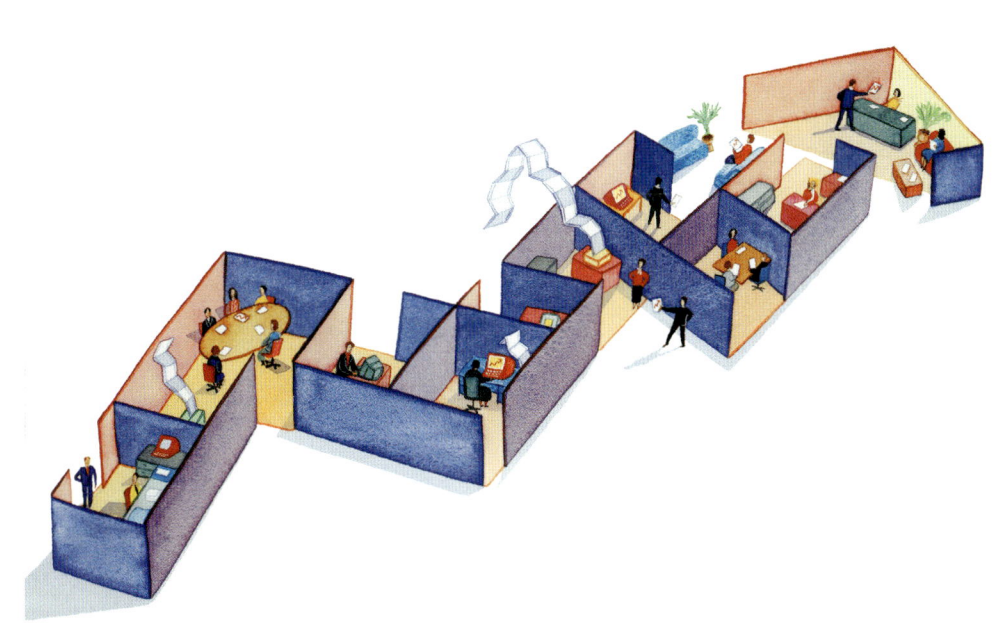

High-impact people and high-performance teams live at the very heart of any company's capacity to thrive in a competitive environment. Many AEC firms have instigated remarkably effective programs to nurture their most valuable asset, people.

High-Impact People   163

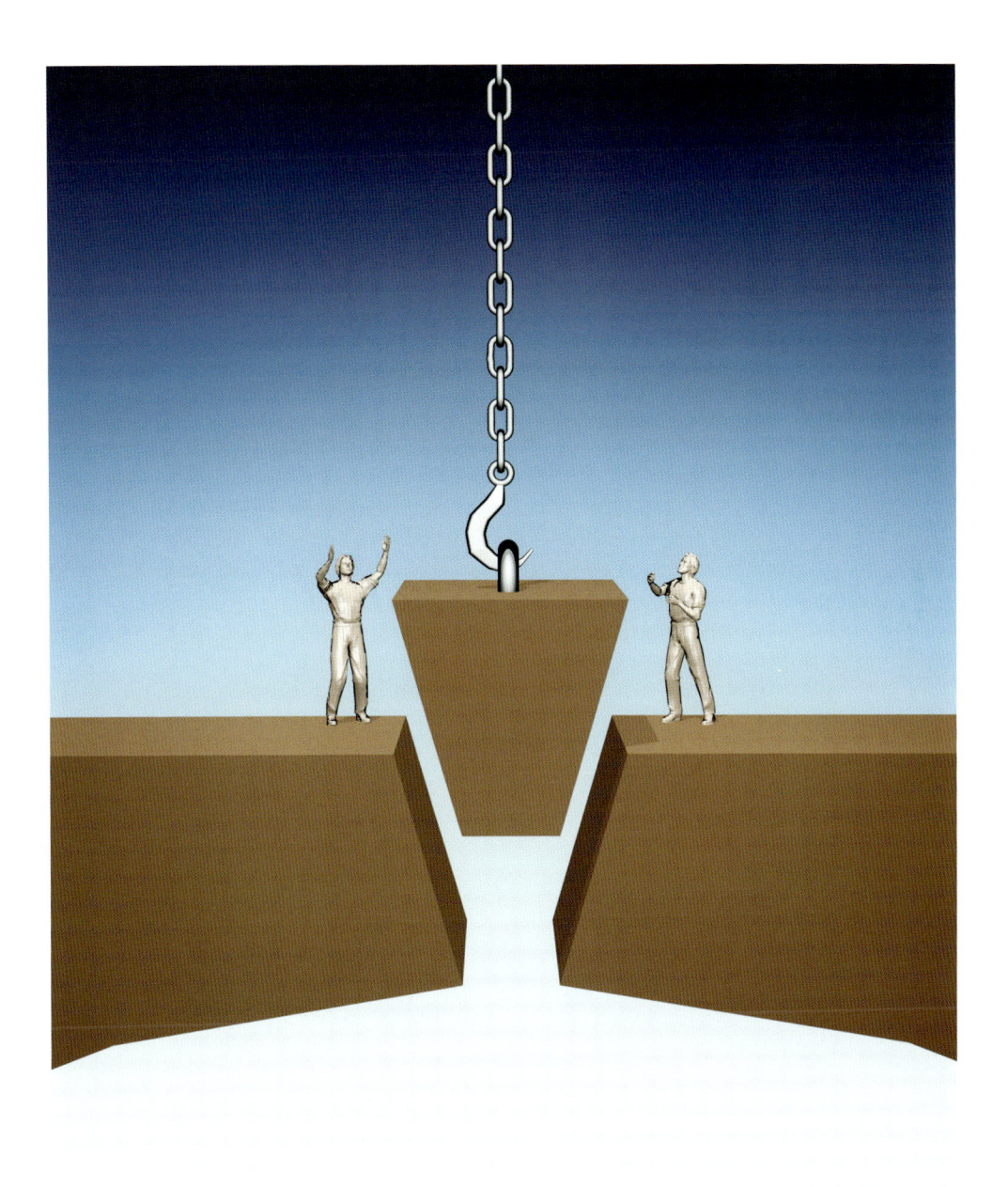

New Strategic Relationships
Any company not actively engaged in exploring new, integrated project relationships has placed itself on the fast track to irrelevancy.

Chapter 8

NEW STRATEGIC RELATIONSHIPS

How are integrated project teams evolving?

How are client relationships evolving?

The first seven chapters of this book could actually be referred to as a primer for the application of current management thinking to competitiveness in the AEC industry. The material contained herein simply takes what we at the Rice University Building Institute have learned from the last three years of interacting with our research partners, filters it through lessons gained from leading management scholars, and points out how architects, engineers, and builders can apply it to become more competitive.

This chapter, however, is quite a different matter. It addresses an issue that has become so pressing as to demand an immediate response from all companies aspiring to remain competitive in the building industry:

Sequential project delivery will no longer be tolerated.

Traditional linear project delivery models are characterized by professionals who execute their responsibilities and then hand off the work to the next experts in the system. Face-to-face collaboration is typically minimal and, during some phases of the work, information flow often spends as much energy running backwards as it does forwards. This approach is rapidly on its way out, as it should be. Any company not actively engaged in exploring new, integrated project relationships has placed itself on the fast track to irrelevancy.

Better processes
Architects, engineers, and builders regularly attack and solve incredibly complicated problems. Most are masters at integrating a wild variety of complex systems into a coherent whole. Why then, have we been so remiss in turning our attention not only to designing and building better buildings, but to designing and building better project delivery processes?

How are integrated project teams evolving?

We are all aware of and familiar with the traditional sequential project team structures that have held sway for the last 150 years. In most industries it is common to observe older, less efficient systems methodically yield to procedures that are simpler, faster and more effective. This gradual improvement process is not going to be allowed in the AEC world. Having interviewed hundreds of major, repetitive buyers of design and building over the past three years, we haven't found a single owner that isn't exploring and demanding new project delivery strategies. This movement toward a more integrated process is not progressing at a civilized, mannerly trot. It is, in fact, a stampede.

Managing Innovation in Construction
Martyn Jones and Mohammed Saad
Thomas Telford, 2003

The next generation of market leaders will have no choice but to lead the way.

> The generic term is integrated project delivery, and what it really means is abandoning the sequential decision-making process and substituting for that a simultaneous process of decision making that involves the end users, the clients, and the contractors at the beginning of the job. And it works wonders. The average savings on projects we've had using this method is $5.4 million. The average savings in time on the schedule is 8.4 months. And every single time we've used the process, the building has been a design award winner.
>
> SCOTT SIMPSON, FAIA, MANAGING DIRECTOR
> KLINGSTUBBINS
> CAMBRIDGE, MASSACHUSETTS

International survey

So, at this point, what have we learned about the imperative to reorganize our project relationships? At the Rice University Building Institute, we decided to initiate the conversation by conducting an internet-based international survey. Even though the survey is scheduled to go active in the US, Britain, Canada, and Australia, at this point we only have access to the US data. Upon completion, in about one year, we will, of course, publish the results.

Our respondents include owners, project/program managers, architects, engineers, construction managers, and major subcontractors.

Innovation in Construction: A European Analysis
Marcela Miozzo and Paul Dewick
Edward Elgar, 2004

Since the concept of integrated project delivery is still in its infancy, we thought it would be instructive to define current attitudes and opinions about the most popular project delivery systems currently in use.

Our survey deals with attitudes concerning the following questions:

1 How do owners select teams?
Owners assemble their teams in a wide variety of ways. Which are most common and which are most effective?

2 Who should manage the process?
Which of the interdisciplinary team members are most qualified to manage the process?

3 How do each of the team members define project success?
Here we want to understand the possibility that one of the team members can have a successful project while another team member experiences major problems. For instance, the construction manager makes a good profit but the engineers lose money or the architects get a design award but the owner is unhappy with the building.

4 How do teams collaborate?
How does each team member perceive the effectiveness of the team's group problem-solving capabilities? We also want to know to what extent each team member feels their input is being valued.

5 What problems occur? What are the most common sources of dissatisfaction
Specifically, we want to know if any team member's dissatisfaction stems from repetitious and predictable events.

6 Do prejudicial attitudes exist about project delivery systems?
We want to know how individual team members feel about selected delivery systems even before the work starts.

Who Should Manage the Process?
Which of the interdisciplinary team members are most qualified to manage the process?

Partnering and Alliancing in Construction Projects
Sally Roe and Jane Jenkins
Sweet & Maxwell, 2003

7 How do you match systems to quality, time, and budget?
Most team members have experienced a variety of project delivery systems. Which system do they believe gives them the best chance to deliver an innovative building? Which is best for delivering on time? Which is best for delivering on budget?

> Our historical success is built on the notion that we deliver as agreed, sometimes at all costs. To attain this well-deserved reputation, we've had to be a pretty tough customer. But what we're seeing in the Australian market in recent years is that's not enough. Our customers are becoming more successful through new alliances, relationships, and collaborative behaviors, which Leighton hasn't traditionally seen as its key performance trait. So, we're actively pursuing new integrated/partnership style project delivery models.
>
> MIKE ROLLO, GENERAL MANAGER, BUSINESS SERVICES
> LEIGHTON CONTRACTORS
> SYDNEY, AUSTRALIA

This survey asks for opinions and attitudes about six typical project delivery systems currently in use around the world. As we all know, the specific definitions of these systems vary widely depending on who you ask. For the purpose of our survey, however, we must all be working from consistent definitions, so we provided the following simple descriptions:

1 Design/bid/build
In this process, the owner selects an architect/engineer to fully design and document the proposed building. When, supposedly, all decisions are made and documented, the owner takes bids from a collection of qualified general contractors and one is selected. The general contractor is then responsible for all methods and costs of construction, including the payment and performance of all subcontractors.

2 Design/build
This is a system in which the owner contracts with a single entity to design and construct the proposed building. This single entity may be an "integrated design/build company," a joint venture

between an A/E firm and a general contractor, or a traditional general contractor that has chosen to specialize in design/build. The design/builder contracts directly with all subcontractors and is a single point of responsibility for all facets of design and construction.

3 Construction manager at risk
This system requires that the owner select the architect/engineer and the construction manager concurrently. All are involved in the design and documentation phases and the construction manager then generates a guaranteed maximum price. The construction manager engages all subcontractors and is ultimately responsible for delivering the project on time and on budget.

4 Construction manager agency
This works similarly to the construction manager at risk system except that the construction manager serves as the owner's agent and works for a fee. The owner contracts directly with all subcontractors, who in this scenario are often referred to as "multiple primes." With this approach, there is usually no guaranteed maximum price and the owner assumes all risk of construction.

> If an owner is willing and interested, he can trade "risk" for lower cost and greater control. When a project exercises the traditional design/bid/build method, the owner buys the management of risk in the General Contractor's bid. CM Agency/multi-prime with full pre-construction represents a shift in the paradigm for any of the traditional delivery methods and puts the Owner more in control of the management of his own project.
>
> STEPHEN MARTIN, PE, DIRECTOR OF PROGRAM MANAGEMENT
> JACOBS
> SACRAMENTO, CALIFORNIA

5 Bridging
The object of this system is to have the design architect/engineer prepare only "design intent" drawings. The design team is free to specify and describe all features they feel are essential to achieve the desired outcomes. The construction manager then

Alliancing
Originating in Australia, this is a team selection and remuneration scheme that can be integrated with several other systems. The idea is to increase team collaboration by creating a bonus system that is shared by all or by none.

coordinates the major subcontractors who generate a guaranteed maximum price and prepare final construction documents.

6 Alliancing

Originating in Australia, this is a team selection and remuneration scheme that can be integrated with several other systems. The idea is to increase team collaboration by creating a bonus system that is shared by all or by none. Obviously, team members should be motivated to catch and repair each other's mistakes. If the project does not achieve its intended benchmarks, no team members receive the bonus payment.

Each of these project delivery systems engages a large, complex cast of professionals and experts. Each functions within its own legal and risk management environment. For the purposes of our survey, it was necessary create these simple descriptions, however, no system is as uncomplicated as we just described it.

Construction Partnering and Integrated Teamworking
Gill Thomas and Mike Thomas
Blackwell Publishing, 2005

In the future, we will need to utilize more innovative and non-traditional approaches to project delivery in order to effectively manage cost, schedule, and quality concerns. I see three innovations which should receive considerable attention: (1) strategic project launch assessment as first step to project implementation, (2) implementation of building information modeling techniques, and (3) implementation of lean process techniques.

JOHN E. KEMPER, CHAIRMAN & CEO

KLMK GROUP

RICHMOND, VIRGINIA

Even though this international survey is ongoing, we can begin to draw some early conclusions:

Infinite variations

Even though this survey instrument provided six delivery options as a basis for our survey, owners and consultants are customizing their processes in an infinite variety of ways in an attempt to find the optimum arrangement for the given circumstances.

> A truly integrated team approach, with the client functioning as an active partner, allowed us to simultaneously create a project budget, project schedule, project program, and 30-acre master plan. Our team literally reinvented not only our work process, but our professional relationships as well.
>
> LAURA STILLMAN, EXECUTIVE VICE PRESIDENT
> FLAD ARCHITECTS
> MADISON, WISCONSIN

Different opinions/different views

Participants in the AEC industry are not starting this conversation with a common descriptive vocabulary. Not only do we repeatedly use the same words to express different ideas, we have vastly different feelings about what are considered standard approaches. Mention the term "design-build" to a large group of AEC people and some will instantly experience a deep warmth in their souls while passionately holding forth on why this is a superior approach. Others will run screaming to the nearest exit.

Owners are more knowledgeable and sophisticated

The vast majority of projects from which we learn were initiated by owners who are experts. They build repetitively and therefore have assembled in-house staffs of experienced professionals to manage the process. Take note of Charles Thomsen's description of "rotation" later in this chapter.

Legal/insurance/bonding issues

Improving any complicated process usually requires retooling an interconnected web of activities, none of which can be adjusted without impacting all others that touch it. When we attempt to evaluate all major moving parts of today's project delivery processes, two issues invariably bubble forth: Revising the legal environment in which this takes place will be difficult for all and the same problem exists for fixing the risk management environment. Later in this chapter, we discuss an "integrated agreement" signed by all major project team members, which may in fact represent the future of project contracting.

Participants in the AEC industry are not starting this conversation with a common descriptive vocabulary. Not only do we repeatedly use the same words to express different ideas, we have vastly different feelings about what are considered standard approaches.

> These emerging integrated delivery processes are proving to be enormously effective at creating efficiencies and minimizing errors during construction. As they are implemented, it is critically important that traditional notions of risk allocations among participants, each of which is a potential legal adversary, change to foster collaboration.
>
> JOHN R. HAWKINS, AIA, ATTORNEY
> PORTER & HEDGES, LLP
> HOUSTON, TEXAS

Who should manage the process?

This particular group of questions has provided an entertaining and predictable series of responses. We are trying to measure how each set of professionals feels about the other's management skills. Remember that, for these inquiries, our respondents have been divided into categories: owners, project managers, architects, engineers, and general contractors.

You'll be shocked to learn that each group thinks they would be best at managing the entire process. That's right. The architects think they should be in charge as do the project managers and general contractors. We then asked: "If you can't be in charge, who would you select?" Here, all participants agreed. They all selected the project managers as most qualified. The most amusing responses came to the last question in which we inquired: "Which group would you least like to see managing the process?" One of the possible responses was "multiple leaders" and it was selected as the worst possible approach by everyone except the general contractors. The general contractors selected the architects as even worse than multiple leaders.

What defines project success?

Once again, with these questions we are trying to delineate divergent attitudes that have probably existed for years among interdisciplinary building team members. This is critical because we believe these very different professional views represent the gremlins responsible for retarding our progress toward more integrated and effective project delivery systems.

Our respondents were given four measures of project success from which to choose: (1) quality of process, (2) quality of building, (3) time and budget, and (4) client satisfaction.

We found, as expected, that owners and project managers defined project success utilizing exactly the same measures. All, of course, felt that time and budget considerations were paramount, but the architects were far more motivated by client satisfaction than the others. Interestingly, the group most concerned with quality of process, by a wide margin, was the general contractors.

Sources of dissatisfaction

Even though most project team members define success by time and money, not meeting those goals were never the greatest source of dissatisfaction. It was always teamwork and communication. It's not the external problems that cause dissatisfaction, such as weather delays, material cost increases, and unexpected change orders. It's the way those disruptions are handled by the team. Project teams characterized by high levels of trust and collaboration seem to work their way through difficult situations with minimum lingering tension.

We learned about this organizational reality years ago. Within most companies, it is not the external stresses that ratchet up frustration. It is the way they are processed internally.

What is each system's strength?

This survey will be executed in four countries: USA, Britain, Canada, and Australia. We began in the US and at this interim point, we only have American data. We are aware, therefore, that these early conclusions are substantially skewed to US practices and are not currently as interesting or useful as when we get data from around the world.

That being the case, our respondents have a great deal of experience with "design-bid-build," "design-build," "construction manager at risk," and "construction manager agency." They have less experience with "bridging" and even less with "alliancing." Keeping this in mind, here is an interim summary of the responses to the question: "Which system gives you the best chance to produce a building?"

- **On time:**
    - architects/engineers—design-bid-build
    - contractors—design-build
    - owners—design-build
- **On budget:**
    - architects/engineers—design-build
    - contractors—design-build
    - owners—alliancing
- **Innovative:**
    - architects/engineers—design-bid-build
- **Building:**
    - contractors—design-build
    - owners—construction manager at risk

This is clearly an interesting divergence of attitude. It speaks volumes about the task faced by tomorrow's market leaders as they attempt to create a genuinely integrated project delivery system.

So what does all this mean to those companies aspiring to be the next generation of AEC market leaders?

Action now
A wait-and-see approach will sink you.

Learn
Like any new compelling professional movement, integrated project delivery is being studied and advanced by a wide variety of organizations. We have collected position papers and studies from the architectural, engineering, and construction professional associations in the US, Britain, Canada, and Australia. Intelligent people are proposing interesting approaches and we must all make ourselves aware of the latest thinking.

Innovate and test
Remember the old saying: "You can always identify the pioneers, they're the ones with the arrows in their backs." Innovation and advancement in project delivery is imperative, but it doesn't have to be foolhardy. Don't proceed without a reasonably comprehensive study of all project delivery advances made in the last two years.

Pockets of innovation

We know that all project delivery strategies involve a complex collection of players critical to the process: owners, project managers, architects, engineers, construction managers, major subcontractors, suppliers, manufacturers, attorneys, etc. In search of the latest thinking, we find innovations emanating from almost every corner. No one group or discipline has a lock on the revolution. Included in this chapter are brief explanations of seven pockets of innovation that we have uncovered in our research on competitiveness. This is not a comprehensive analysis of all project delivery advancements, but rather a smattering of brief stories worthy of attention.

Pockets of Innovation
In search of the latest thinking, we find innovations emanating from almost every corner. No one group or discipline has a lock on the revolution.

Pockets of innovation

Scott Simpson, FAIA, Senior Principal for Strategic Growth
KlingStubbins

William A. Daigneau, Vice President of Operations and Facilities Management
University of Texas M D Anderson Cancer Center

Jeffrey C. Hines, President
Hines Interests

Barbara White Bryson, Associate Vice President, Rice University

Charles B. Thomsen, Former Chairman of CD/I, Advisory Director, Parsons

Richard A. Morris, Executive Vice President
Smith Seckman Reid, Inc.

Glenn Ballard, Research Director, Project Production Systems Laboratory
University of California, Berkeley

David Chambers, Chief Architect
Sutter Health, Sacramento, California

William Lichtig, Outside Counsel
Sutter Health

High Quality Low Cost
"What we are trying to create is faster decision making, higher levels of decision making, and better outcomes."
Scott Simpson, KlingStubbins

Higher quality, lower cost

Scott Simpson, FAIA, Managing Director • KlingStubbins • Boston, Massachusetts

What we are trying to create is a higher level of engagement early in the process among the owner, designer, and contractor. This leads to faster, better decision making and better overall outcomes. For example, we're currently doing a major research center in New York: three glass towers comprising nearly 1 million square feet. Using the principles of integrated project delivery, we brought in the fabricator and installer to work alongside the architect to design the curtainwall. Many of their suggestions were adopted and in fact some of their technical drawings were embedded right into the CDs. The result was a much better design, and the price dropped from about $125 per square foot to $104 per square foot. We expect very few problems in the field (and no RFIs!) because of this close collaboration. It's a win–win for all concerned.

Agile development for technology-impacted industries

William A. Daigneau, Vice President of Operations and Facilities Management • University of Texas M D Anderson Cancer Center • Houston, Texas

---

Agility is a common goal for most industries that want viable products delivered to the marketplace faster. Proven successful concepts include flexibility, concurrent processing, synchronization, and information management.

When such concepts are applied to construction projects delivery time is reduced 30%–50% from conception to completion. The following four-step process can increase a facilities project's value to the institution through reduced interest costs, faster space availability, and flexibility to adapt to the user's ever-changing needs.

Step 1: Set building location, size, and intended use
Once a preliminary building need and scope is established, a construction budget can be generated using actual costs of similar building types. This becomes the benchmark total project cost. At this point it is essential to gather only information required to gain project approval and funding. Actual building occupants and interior details will be programmed later in the process.

Step 2: Select project delivery method and organize team
Next the design/engineering/construction team is contracted, and the owner's core team of users, operators, and regulatory experts is assembled. All relevant constituents meet weekly, utilizing a formalized format, to insure timely management decisions.

Step 3: Establish flexible floorplate and basis of design
With the conceptual use of the proposed building known, design can proceed sufficiently to define the structure, exterior, and overall mechanical, electrical, and plumbing systems. Since the final occupants of the building are yet to be defined, we plan to accommodate the highest intensity use. This allows for a generic building capable of handling a wide variety of evolving uses.

Step 4: Initiate detailed programming, design, and construction
With the building's basic systems established, the structural frame, exterior skin, and site utilities can commence construction. Next, detailed programming is initiated that will define final fit-out and finishes. The initiation of the final programming and design sequence is tied directly to the last possible moment information is required by the construction schedule. Delaying these decisions to the last moment helps reduce the project's exposure to changes in technology and work processes that inevitably affect building function and its final cost.

Focused, market-driven design

Jeffrey C. Hines, President • Hines Interests • Houston, Texas

How much time do architects and engineers spend coming up with interesting ideas that eventually get rejected? The Hines organization attempts to reduce that wasted energy by creating a unique partnership with the architects the firm chooses. And this is usually accomplished without sacrificing architectural or functional integrity.

Our unique approach to design is at the heart of our capacity to produce, on a global basis, functional, elegant, sustainable, market-driven buildings. Our process begins with a central resource group called Hines Conceptual Construction. More than a pre-construction department, this group uses 50 years of best practices, as well as mistakes from which we have learned, to inform each new project we tackle.

For every project, we assemble an integrated team of architects, engineers, contractors, major subcontractors and marketing specialists. We prepare a benchmarking document that analyzes currently available competitive buildings and defines standards in a given market. The entire team then discusses how we want our project to respond to market forces. We review all system and technology alternatives, such as sustainable features, electricity consumption, air filtration, security ... everything.

In some cases, the team instantly knows which way we want to go. For others, further study is required. At the end of the conceptual process, the team has a clear mandate: a statement of exactly what we want to build, why we want to build it, how we expect the project to perform financially and what its short- and long-term effects on the community and the environment will be.

The design team, rather than feeling constrained, typically thinks this level of focus keep them from wasting design time on ideas that would most likely be rejected. Also empowered by this level of focus is the marketing group. Since they are involved early, they can communicate better to brokers and prospective tenants.

Planning in advance to this extent also pays another important dividend for the firm: costly change orders are rare, and our investors appreciate that a great deal.

Collapse the information time line

Barbara White Bryson, Associate Vice President • Rice University • Houston, Texas

The single most inefficient aspect of the design and construction process is communication. The simple fact is that, in the design-bid-build system, architects and engineers cannot access the information they need to develop the best designs until long after the design is substantially complete. By then, time is lost, efficiency is lost, and the opportunity to make great strategic decisions is lost. It is the job of the entire building team of the 21st century to collapse the information time line, to fold the right information into the beginning of the design process where it can be most effective.

This is already being done by pulling subcontractors into pre-design while creating new strategic alliances by architects and manufacturers. The true solution to this problem may be the reinvention of the AEC industry itself.

We need to forget the notion that fast means lower quality or higher cost. No other industry believes this is true. "Faster" can mean better and more beautiful. Time can be a design constraint as compelling and energizing as a distant cliff view or a tight urban corner. Time is simply a design challenge, and when design teams figure that out, their power will be greatly enhanced.

Moving project workflow into program workflow

Charles B. Thomsen, FAIA, FCMAA and Former Chairman, 3D/I and Advisory Director, Parsons • Houston, Texas

Most of the buildings, these days, are built by owners who are serial builders. The projects are part of a program. Program managers have an important duty that project managers don't have. They can examine the projects to find similarities, then focus on ways to improve them. Sometimes the results can produce enormous benefits in time, cost, and quality across the entire program.

Rotation is a business term used to describe the process of turning a custom, project-oriented activity into a continual, program-oriented standard and improving that standard to increase productivity. We use the term to describe moving unique activities in a project workflow to a continual activity in the program workflow.

The degree of rotation that can be achieved in a program is a function of the number of projects, the similarity of the projects, and the authority of the program manager to enforce standards and push improvement. Rotation requires: (1) analyzing the similarities that exist among projects, (2) choosing those similarities that are the most repetitious and offer the greatest possibilities for improvement and standardization, and (3) focusing on ways to improve these standards at the program level.

Of course, every project has unique parts. However, within a program there are always similarities that offer an opportunity for continuous improvement. The challenge is to reduce the casual, arbitrary uniqueness to only those areas that have a genuine unique requirement. Often when we talk about rotation, people respond that it won't work for them because they must build different kinds of buildings. So, let's get this notion out of the way. Capturing the benefits of rotation does not require designing a prototype and plunking down cookie cutter replications.

Whole team involvement and BIM

Richard A. Morris, Executive Vice President • Smith Seckman Reid, Inc. • Houston, Texas

Truly integrating today's design and construction services is the focus and groundswell of the entire AEC community. Over many decades, we have made little progress in re-tooling our traditional linear delivery model of design, bid, build. One approach to improving the traditional delivery model is integrating the efforts of the design and construction community through the use of building information modeling (BIM). While the process has been in existence for over two decades in the industrial community, the AEC industry is relatively late to the game. Clearly, today, most design and construction partners see the advantage, and are promoting whole team involvement.

The idea is to bring collaboration to a higher level than afforded by the traditional linear thought process. This should allow the owner to make more informed decisions and create savings in project delivery time resulting in overall project cost savings.

In multiple projects, we have found the 3D process to be an enhancement in design. It assists our firm in making better use of space both horizontally and vertically, enabling our design team to optimize engineering spaces and improving the overall building utilization. We have also experienced enhancements in our quality assurance plans during the design process, allowing the team to resolve issues faster and better during the construction phase.

The ultimate "integrator" lies in involving the construction team during the entire design process and sharing the 3D model to create better awareness of the design intent. This provides increased confidence on the part of our construction partners, allowing them to provide tighter pricing and bids. Since the contractors have been involved in the design process and are fully aware of its intent, they can resolve many issues without further involvement of the design team, thereby reducing RFIs and change orders as well as the administrative effort required to create and monitor the paper trail. While software inter-operability problems currently make it difficult to completely share the 3D model with all construction partners, the AEC community is making rapid progress.

In a traditionally linear delivery model, the decades-old familiar process is quickly evolving to a much more integrated process including all of the stakeholders with the prime benefactor being the owner.

"Over many decades, we have made little progress in re-tooling our traditional linear delivery model of design, bid, build. One approach to improving the traditional delivery model is integrating the efforts of the design and construction community through the use of building information modeling (BIM)."
Richard Morris, Smith Seckman Reid

New Strategic Relationships

Development of the Integrated Agreement for Lean Project Delivery

Glenn Ballard, Research Director, Project Production Systems Laboratory • University of California, Berkeley

David Chambers, Chief Architect • Sutter Health • Sacramento, California

William Lichtig, Outside Counsel • Sutter Health • Sacramento, California

In order to fully embrace the concept of lean project delivery, Sutter Health determined it should develop a relational contract—an agreement that would be signed by the architect, construction manager/general contractor, and owner—and would describe how they were to relate throughout the life of the project.

The Integrated Agreement is a single document that is signed by the architect, the CM/GC and owner. It is not a design/build contract, where one entity takes total responsibility for all aspects of project delivery. Instead, the agreement describes the relationships that are established among each of the members of the integrated project delivery (IPD) team.

Rather than being conceived as a "three-legged stool," this primary relationship is depicted as three overlapping circles. The project representatives for each of these entities form the "core group." This group has primary responsibility for the selection of the rest of the IPD team and for management and operation of the project. Most major project-related decisions are to be made by consensus of the core group. Only in the event of impasse does resolution transfer to the owner.

The parties recognize that each of their opportunities to succeed on the project is directly tied to the performance of the other participants.

A final word

The AEC industry will never realize its full economic potential until we reposition our efforts in the hearts and minds of our clients. As long as they see us as people who only design and build buildings, our value to the business and cultural communities will stagnate if it hasn't already. It is imperative that we be perceived as people who utilize our knowledge of the built environment to address and improve business and cultural problems.

Our industry is not just about the buildings.

As long as clients see us as people who only design and build buildings, our value to the business and cultural communities will stagnate if it hasn't already. It is imperative that we be perceived as people who utilize our knowledge of the built environment to address and improve business and cultural problems.

How many new graduates of architectural, engineering, or construction management schools can sit down across a conference table from the CEO or CFO of a major corporation and have an intelligent conversation about the corporation's diverse collection of business strategies? And next, explain how the built environment can positively impact the success of those strategies? Today, not many. Tomorrow, hopefully all of them.

QUOTED INDUSTRY LEADERS

Often academic and industry leaders go about their business with little meaningful interaction. When effective collaboration occurs, however, it can create valuable and powerful insights. We are indebted to the following executives for sharing their wisdom.

BALLARD, Glenn
Research Director – Project Production Systems Laboratory
University of California, Berkeley
Berkeley, California

BEARDEN, James
CEO
Gresham, Smith & Partners
Nashville, Tennessee

BLACK, Tim
Director
BKK Architects
Melbourne, Australia

BOWKER, Margaret
Assistant VP
JE Dunn
Kansas City, Missouri

BRENNAN, Lee
Principal
Cannon Design
Los Angeles, California

BRYSON, Barbara White
Associate VP
Rice University
Houston, Texas

BUTTERFIELD, Leslie
CEO
McLachlan Lister Pty Ltd
Sydney, Australia

CARETSKY, William
Senior VP
Syska & Hennessy
New York, New York

CHAMBERS, David
Director
Sutter Health
Sacramento, California

CLEMENTS, Robert
Senior VP
Jacobs Engineering Group Inc.
Pasadena, California

DAIGNEAU, William
VP, Operations & Facilities Management
UT MD Anderson Cancer Center
Houston, Texas

DRUMMOND, Peter
CEO
Building Design Partnership
London, England

FIRLOTTE, Frederick
President & CEO
Golder Associate
Montreal, Canada

FORSYTHE, Stephanie
Principal
Molo
Vancouver, Canada

FUKSAS, Massimilliano
Principal
Massimilliano Architects
Rome, Italy

GRAHAM, Stuart
President & CEO
Skanska
Solna, Sweden

GRECO, Chuck
President & CEO
Linbeck Group, LP
Houston, Texas

HARRISON, Doug
Manager of Administration
Stuart Olson Construction
Calgary, Canada

HAWKINS, John R.
AIA Attorney
Porter & Hedges, LLP
Houston, Texas

HAWKINS, Ralph
President & CEO
HKS
Dallas, Texas

HEALD, Peter
Executive Vice President
CGI
Los Angeles, California

HILDEBRANDT, Melvin
Former President
Linbeck
Houston, Texas

HINES, Jeffrey C.
President
Hines Interests
Houston, Texas

HODKINSON, Gregory
Chairman
ARUP Americas
New York, New York

HOLMES, Wendell
Regional CEO
Gilbane
Providence, Rhode Island

HUGHES, Greg
Principal
Perkins & Will
Houston, Texas

JOHNSON, Robert
Executive Director, CRS Center, College of Architecture
Texas A&M University
College Station, Texas

JONASSEN, Jim
Managing Principal
NBBJ
Seattle, Washington

JOSAL, Lance
Senior VP
RTKL Associates
Chicago, Illinois

KEENBERG, Ron
Chief Architect
IKOY
Ottawa, Canada

KELLY, David
Managing Director
Gleeds
Sydney, Australia

KEMPER, John E.
Chairman & CEO
KLMK Group
Richmond, Virginia

KIRKSEY, John
President
Kirksey Architecture
Houston, Texas

KOTLER, Phillip
Expert & Author
Kotler Marketing Group
Washington, DC

KUHNE, Eric
Principal
Eric R. Kuhne & Associates
London, England

LATHAM, Sir Michael
Deputy Chairman
Willmott Dixon
London, England

LEA, Jerrold
Senior VP
Hines
Houston, Texas

LEVINE, Robert D.
Senior Vice President
Turner Construction
New York, New York

LICHTIG, William
Outside Counsel
Sutter Health
Sacramento, California

LINBECK, Leo
Chairman
Linbeck
Houston, Texas

MANASC, Vivian
Principal
Manasc Architects
Edmonton, Canada

MARTIN, Stephen
Director of Program Management
Jacobs
Sacramento, California

MESSER, Raymond
President & Chairman
Walter P. Moore
Houston, Texas

MILLER, Stephen
Society of Competitive Intelligence Professionals
Alexandria, Virginia

MINTZBERG, Henry
Associate Professor
McGill University
Montreal, Canada

MORRIS, Andrew
Senior Director
Rogers Stirk Harbour & Partners
London, England

MORRIS, Richard A.
Executive Vice President
Smith Seckman Reid, Inc.
Houston, Texas

NOOK, Greg
Executive VP
JE Dunn
Kansas City, Missouri

OGILVY, David
(1911–1999)
Founder, Ogilvy & Mather

PECK, Chris
VP
McCarthy Building
St. Louis, Missouri

PENLAND, Will
Senior Managing Director
CB Richard Ellis
Houston, Texas

PRINGLE, Jack
President, RIBA ('05–'07)
Director, Pringle Brandon Consulting
London, England

RATCLIFF, Kit
Principal
Ratcliff Architects
San Francisco, California

REED, Scott, RAIA
Western Region Market and Practice Leader
Cannon Design
Los Angeles, California

ROBSON, Thomas
Senior VP
HOK
St. Louis, Missouri

ROGERS, Peter
Director
Stanhope PLC
London, England

ROLLO, Mike
General Manager
Leighton Contractors
Sydney, Australia

SCHMITT, Don
Principal
Diamond + Schmitt Architects
Toronto, Canada

SHEPHERD, Nick
Managing Partner
Drivers Jonas
London, England

SIMPSON, Scott
Senior Principal
KlingStubbins
Boston, Massachusetts

SKERMAN, Ben
Engineer
Worley Parsons
Queensland, Australia

STEWART, R.K.
Principal, Gensler & President, AIA (2007)
San Francisco, California

STILLMAN, Laura
Executive VP
Flad Associates Architects
Gainesville, Florida

STOWKOWY, Al
President & COO
Stuart Olson
Calgary, Alberta Canada

TELLEPSEN, Howard
Chairman
Tellepsen Builders
Houston, Texas

THOMSEN, Charles
Former Chairman, 3DI & Advisory Director, Parsons
Houston, Texas

TRIONE, Gerry
Executive VP
CB Richard Ellis
Houston, Texas

TZANNES, Alec
President, RAIA (2007) & Founder of Tzannes Associates
Sydney, Australia

WYATT, Scott
Managing Partner and Principal
NBBJ
Seattle, Washington

INDEX

advancement, clear path to, high-impact people 156–7
AEC industry 2
agile development for technology-impacted industries 182
    establish floorplate and basis of design 182
    initiate detailed programming, design and construction 182
    select project delivery method, organize team 182
    set building location, size and intended use 182
agility, a function of culture 56
alliancing 74, *172*, 173
American Institute of Architects (AIA) 27
architects viii
    next generation, mission of 141
architecture, restoration of storytelling qualities 11
The Art of War Sun Tzu 106
Arup Americas 99, 100, 157, 192
    Fellowship Program 160
Association of General Contractors (AGC) 27

Ballard, Glenn (RD – Project Production Systems Lab., University of California) 180, 188, 191
Bearden, James (CEO, Gresham, Smith & Partners) 140, 191
BIM *see* building information modeling (BIM)
BKK Architects (Melbourne, Australia) 90, 191
Black, Tim (Director, BKK Architects) 90, 191
Bowker, Margaret (Assistant VP, JE Dunn) 132, 191
brand 92
brand alignment 92
brand credibility 92
brand equity 93
brand essence
    focus on emotion 94
    meaning test 94
brand experience 94
brand extension 96
brand gap 97
brand metrics 97
brand pushback 97
brand strategy 98
brand touchpoints 98–9, *98*
branding 88
    and dissemination, creative 75
    intelligence gathering externally focused 103, *104*
    role of competitive intelligence in 102–9
        better decisions 105–6

        indicators of competitive success 102–3
        market leaders generate information in four areas 107, 107–8
        perceptive questions 106
        quality data 106
    utilized by today's market leaders 99–102
    value integrated branding events 45–6
    see also persistent branding
branding programs, getting started 113–18
    create internal branding team 113
    dissemination of your brand 117–18
        dissemination via your staff 118
        get started with even a modest budget 117
    measure impact of existing brand 113–15
        define the perception delta 116
        design relevant survey techniques *114*, 115
        devise a branding action strategy 116
        evaluate and adjust 118
        execute branding identity investigation 115–16
brands, essence or slogans 95
Brasfield & Gorrie 95
Brennan, Lee (Principal, Cannon Design) 118, 156, 191
Bryson, Barbara W. (Associate VP, Rice University) 180, 191
    Collapse the Information time line 184
Building Design Partnership (London, England) 52–3, 191
building industry, information lacking 143
building information modeling (BIM) 186, *187*
    3D process, an enhancement in design 186
buildings
    infiltrated by behaviors and attitudes viii
    lack of pre-occupancy studies viii
    making of a competitive business viii–ix
    produced on time, which system gives the best chance 177–8
    study of buildings in use viii
built environment, shaping of xi–xiv
    approach xii
    research xi–xii
    surveys, interviews, focus groups xii
Butterfield, Leslie (CEO – McLachlan Lister Pty Ltd) 24, 56, 191

Cannon Design (Los Angeles) 95, 108, 118, 191, 194
    Yazdani Studio 156
Caretsky, William (Senior VP, Syska & Hennessy) 191
    on personal attention 118

Caretsky, William (Senior VP, Syska & Hennessy)
    cultivating high-impact people 157
category ownership 6, *61*, 62–88
  and the Heisman Trophy! 62–3
  meaning of 63–4
    achieving a level of prominence 63
    common knowledge 64
    devise a new market segment 64
    modify existing market segments 64
  today's category owners, position achievement 65–80
  typical categories 63
Caudill, William (Joint Founder of CRS) 78, 135
CB Richard Ellis (Houston) 193, 194
  in talks with Trione & Gordon 72, 74
CGI Development (Los Angeles) 45, 192
Chambers, David (Director/Chief Architect, Sutter Health) 27, 31, 180, 188, 191
Choquette, Paul (CEO, Gilbane) 39
Clements, Robert (Senior VP, Jacobs Engineering Group Inc.) 14, 34, 191
collaboration, for mutual achievement 30
collaborative process, downside 17
competitive advantage 146–7
  and less then acceptable results 146
competitive culture 7, 124–8
competitive focus 6, *47*, 48–60
  the changing game 49–50
  what is critical about it 50–4
    effective planning 50, *51*
    need to recognizing fads 50–1
  what is wrong with current planning 54–6
    based on common data 54
    culture not suitably agile 56
    doesn't have the right people 55–6
    inadequate focus 56
    not integrated into group consciousness 55
    not translated into measurable missions and goals 55
    process not iterative 55
    takes too long 54–5
    uninspiring 55
competitive focus statements 50
competitive intelligence 7, 110–12
  market leaders generating information 107–8
    centers of influence 107
    competitive selling techniques 107
    competitors 107
    potential clients 107
  role in branding 102–9
competitive intelligence programs 110
  initial steps 112
  routine review and adjustment 112
Competitive Planning at BDP 52–3
  financial objectives and strategic imperatives, monitored and assessed 53
  plan will be changed well before five years 53
  preparation of Five Year Plan 52
  reworking of BDP's core positioning logic, design integrity 52–3
competitive selling 26, 130–9
  important sales attributes 136
  typical selling errors 139
  what do potential clients respond to 134–9
    apparent understanding of considered product 134
    company brand 134
    comprehensive response to entire proposal process 135
    credentials 134
    fees and pricing 138–9
    innovative approaches 134–5
    likeability of sales team 137–8
    responsiveness to client's emotional drivers 135–7
competitive selling techniques 107
competitive strategic planning 48, 57–60
  compelling reason for failure of 48–9
  degrading into a rigid affair 57
  don't ask strategy process to give specific course 59
  old business planning model becoming less relevant 58
  separate process from traditional business planning 59
  use some existing and prospective clients 59
  used by last generation to predict the future 58
competitiveness xi–xii
  increased, and being vision-driven 23–7
corporate stasis, impossible to achieve 122
corporate vision 10
credential differentiation 50
CRS Center (College of Architecture, Texas A & M University) 78, 135, 192
  research and publishing strategy 78
cultural blending 80, *81*
customer intimacy 7, 75
  role played by 82–4, *85*
  techniques that work 83–4
    overcoming attitudes prevalent in upper management 84

Daigneau, William (VP, Operations & Facilities Management, Anderson Cancer Center) 180, 182, 191
design integrity 52–3
design/bid/build system 170
  and collapsing the information time line 184
development and exploitation of new categories opportunity mining
  engage with the corporate world 86
  partnership with best clients *87*
Diamond + Schmitt Architects (Toronto) 194
  and their value system 33–4

Drivers Jonas (London, England) 68, 83, 95, 194
   creation of DJ Sport 67
   development of two sports centers in northern England 66–7
Drummond, Peter (CEO, Building Design Partnership) 52–3, 191
Dunn, JE *see* JE Dunn (Kansas City)

egowranglers 24
emotional intelligence 150
Eric R. Kuhne & Associates 11, 27, 193
execution 146–7
executive overview 2–8
   important questions 4–8
      how to define market prominence 3–4
      next generation market leaders, defining performance characteristics 4–8
expectation management 97

Firlotte, Frederick (President and CEO, Golder Associate) 91, 147, 191
FKP Architects 95
Flad Architects (Madison, Wisconsin) 174, 194
Forsyth, Stephanie (Molo) 54, 191
Fuksas, Massimiliano (Principal, Massimiliano Architects) 15, 192

Galison, Peter
   analogies from the new physics viii
   on success of new physics and its players vii
General Motors, beginning of inexorable decline 12
generic service differentiation 50
Gensler (San Francisco) 95, 143, 194
gifted talkers vs. active listeners 18
Gilbane (Providence, Rhode Island) 192
   evaluation and reward through customer satisfaction 39
   values, six categories emphasized 39
Gilbane University (founded 2000) 39
Gilbane, William J. (President, Gilbane) 39
Gleeds (Sydney, Aus.) 148, 192
globalization 80, 91
Golder Associates (Montreal) 91, 95, 147, 191
Good to Great, Jim Collins 153–4
Gordon, Charles (Part Owner, Trione & Gordon) 70–1
Graham, Stuart (President and CEO, Skanska) 35, 56, 192
Grecko, Chuck (President and CEO, Linbeck Group) 13, 35, 192
Gresham, Smith & Partners (Nashville) 140, 191
Group Identification Theory 36, *37*

Harrison, Doug (Manager of Administration, Stuart Olson Construction) 154, 192
Hawkins, John R. AIA Attorney (Porter & Hedges, LLP) 176, 192

Hawkins, Ralph (President and CEO, HKS) 60, 192
Heald, Peter (Executive VP, CGI Development Services) 45, 192
high-impact people 7, *145*, 146–63
   common causes of losses 163
   creating a new venture from scratch 148–50
      development strategy 148
      sold for large profit 149
   how to keep them 160–3
      career engagement 161
      emotional engagement 161
      flexibility 163
      keep them happy 160
      path to ownership *162*
      personal engagement 160
      recognition 161
      status 161
      support 161
   integration into the work process 153–9
      beware of misdiagnoses 159
      evaluate them on teamwork 158–9, *158*
      focus them 156–7
      identify them 155
      integrate into existing teams with care 154–5
      position them (need for effective compromise) 155–6
      recognize them 157
   in the knowledge delivery business 147, 152
   powerful professional skills and good interpersonal relationships 153
   who they are 147–53
      emotionally intelligent 150
      have superior knowledge and can deliver 152–3
      high capacity for self-criticism 153
      need rapid feedback 153
      seek responsibility 153
      self-aware 151
      self-motivated 151
Hildebrandt, Melvin (Former President, Linbeck) 70, 159, 192
Hines Interests (Houston) 33, 180, 183, 192, 193
   dividends of planning in advance 183
   Hines Conceptual Construction 183
   assembly and mandating of a team 183
Hines, Jeffrey C. (President, Hines Interests) 180, 183, 192
HKS 192
   three-year competitive plan revised annually 60
Hodkinson, Gregory (Chairman, Arup Americas) 99, 100, 157, 160, 192
HOK (St. Louis) 95, 194
   challenges facing healthcare practice 80
   partnering with Skanska 80
Holmes, Wendell (Regional CEO, Gilbane) 39, 192
Hughes, Greg (Principal, Perkins and Will) 12, 192
Hunt Construction Group 95

Index   197

IKOY (Ottawa) 16, 192
individuals, decision to accept/reject group values 36
information gathering techniques 115
information, publicly available 108
innovation, in small, risk-accepting firms 70, *73*
Integrated Agreement 174
   for Lean Project Delivery, Development of 188, *189*
integrated delivery processes 176
integrated project delivery 167, 181
   internet-based international survey 167–73
      common sources of dissatisfaction 168
      matching systems to quality, time and budget 170
      prejudicial attitudes to 168
      problems occurring and common sources of dissatisfaction 168
      process management 168, *169*
      project success, team members' definitions 168
      respondents 168
      team collaboration 168
      team selection 168
   opinions/attitudes about six typical systems 170–3
      CM Agency/multi-prime 171
      alliancing *172*, 173
      bridging 171, 173
      construction manager (CM) agency 171
      construction manager (CM) at risk 171
      design/bid/build 170
      design/build 170–1, 174
integrated project delivery team 188
interdisciplinary collaboration 8

Jacobs Engineering Group Inc. (Pasadena) 14, 34, 191
Jacobs (Sacramento) 171, 193
JE Dunn (Kansas City) 132, 134, 191, 193
Johnson, Dr Robert (Executive Director, CRS Center) 78, 192
Jonassen, Jim (Managing Principal, NBBJ) 18–19, 192
Josal, Lance K, ALA (Senior VP, RTKL Associates) 40–1, 192

Keenberg, Ron (Chief Architect IKOY) 16, 192
Kelly, David (MD, Gleeds Australia) 148, 192
Kemper, John E. (Chairman and CEO, KLMK Group) 173, 192
key employees 15, 16, 59
   and integration techniques 32–3
   in vision-driven companies 24
Kirksey Architecture (Houston) 66, 192
Kirksey, John (President, Kirksey Architecture) 66, 192
Klingstubbins (Boston) 129, 136, 167, 180, 181, *181*, 194

KLMK Group (Richmond, Virginia) 173, 192
knowledge, why giving away is a good idea 75–82
   deeper specialization 80
   don't tell them everything 77
   lessons learned about professional publishing 77
   paths that are overly worn 81–2
   research and publishing strategy 78–80
   risk management 82
   use compelling headlines 77
Kotler Marketing Group (Washington DC) 193
Kotler, Phillip (Expert and Author, Kotler Marketing Group) 86, 193
Kuhne, Eric R. 11, 27, 193
   working with Civic Arts in London 55

lagger problem 21–3
   active support (response – reward them) 23
   go along (and response) 23
   ignore (and response) 22
   overt opposition (and response) 23
   sabotage (and response) 22
   wait and see (and response) 23
Latham, Sir Michael (Deputy Chairman, Willmott Dixon Construction) 50, 193
Lea, Jerrold P. (Senior VP, Hines) 33, 193
leaders, effective 11
leaders, next generation, how will they brand 110–18
   alignment with business strategy 110
   better competitive intelligence 110–12
   getting a branding program started 113–18
   objectively measure outcomes 112
   purposeful vs. intuitive 110
   routine review and adjustment 112
Lederman, Leon 3
Leighton Contractors (Sydney, Australia) 74, 170, 194
Levine, Robert D. (Senior VP, Turner Construction) 155, 193
Lichtig, William (Outside Counsel, Sutter Health) 180, 188, 193
Linbeck (Group), LP (Houston) 13, 35, 159, 192, 193
   pursue relationships not projects 69–70
   and TeamBuild 70
Linbeck, Leo (Chairman, Linbeck) 13, 69, 193

McCarthy Building Companies (St. Louis) 16, 193
McGill University (Montreal, Canada) 60, 193
McLachlan Lister Pty Ltd (Sydney, Australia) 24, 56, 191
management, of high achievers 151
Manasc Architects (Edmonton, Canada) 26, 193
Manasc, Vivian (Principal, Manasc Architects) 26, 193
market dominance 3
market leaders
   dealing with values today 31–42

concerns of today's market leaders 31
development 32, 33–4
evaluation 33
evaluation and reward 39–40
integration 32–3, 34–8
reward 44
summary 41
generate information in four areas 107
historically concern themselves with four things 49
use of personal networking 108, *109*
utilization of branding 99–102
Arup 100
Stuart Olsen 100–2
market leaders, next generation 42
measuring critical behavioral and attitudinal variables *44*
performance characteristics 4–8
use of competitive strategic planning to gain market prominence 57–60
utilizing values to be more competitive 42–6
integrated/interdisciplinary team value systems 44–5
project team value systems 43–4
public status programs 43
recognition programs 43
value integrated branding events 45–6
values champion programs 45
values discovery programs 43
values evaluation programs 43
market prominence
defining of 3–4
name and expertise recognition 4
preferential competitive treatment 4
premium pricing 4
prestige among employees 4
and marketing breakthroughs 6
will be achieved using competitive strategic planning 57–60
market segment redefinition 64, 65
apply an existing service in a new sphere 66–9
deeper specialization 80
define process as a category 69–70
pursue relationships not projects 69–70
globalization 80
identify an underserved market segment 70–1
Trione & Gordon 70–1
joint ventures between innovators and scalers 72–4
budget enough time 74
define performance metrics clearly for all 74
expect defectors 74
expect temporary loss of productivity 74
research and publish 75, *76*, 77–80
communication on a budget 79
importance of differentiation 78
intersection of three concepts 75

research without separate funding 79
market segment specialization 91
marketing breakthroughs 6, *121*, 122–43
getting started 125–8
assign responsibility to senior management 126
begin at the top 126
evaluate performance 128
hold regular progress meetings 126
it's everyone's responsibility 126
make expectations clear 126
program about gathering market intelligence 126
remunerate achievers 128
new business development activities 123
role of competitive culture 124–8
advance competitive position of the company 124
gathering leads, intelligence and opportunities 125
role of competitive selling 130–9
attitude in dominant selling 130
giving out wrong information 130, 132, *133*
what do potential clients respond to 134–9
what constitutes a breakthrough 123–4
capturing new work 123
marketing, role played by 128–30
being proactive 129
first hints of new project opportunities 129
personal networks 129
what is required to get on the short list 130
Martin, Stephen, PE (Director of Program Management, Jacobs) 171, 193
Massimiliano Architects (Rome, Italy) 15, 192
MD Anderson Cancer Center (Houston) 180, 182, 191
Messer, Raymond P.E. (President and Chairman, Walter P. Moore) 69, 193
Miller, Stephen (Society of Competitive Intelligence Professionals) 105, 193
Mintzberg, Henry (Associate Professor McGill University) 193
planning creates plans not strategy 60
Minute Maid Park, Houston 68
Molo (Vancouver) 54, 191
Morris, Andrew (Senior Director, Rogers Stirk Harbour + Partners) 109, 193
Morris, Richard A. (Executive VP, Smith Seckman Reid Inc.) 150, 151, 180, 193
on keeping high-impact people 161
Whole team involvement and BIM 186, *187*
motivators, non-financial, for high-impact people 157

NBBJ (Seattle) 95, 192, 194
Developing Our Company Vision 18–19

NBBJ (Seattle) (*continued*)
    allow entire group to shape the firm 19
    Change Design 18–19
    creating a compelling vision 18
    process as important as outcome 19
new strategic relationships *165*, 166–89
  better processes 166
  how integrated project teams evolve 167–78
    action now 178
    different opinions/different views 174
    infinite variations 173
    innovate and test 178
    international survey 167–73
    learn 178
    legal/insurance/bonding issues 174
    owners more knowledgeable and sophisticated 174
    sources of dissatisfaction 177
    what defines project success 176–7
    what is each system's strength 177–8
    who should manage the process 176
  pockets of innovation 178–89, *179*
    agile development for technology-impacted industries 182
    collapse the information time line 184
    development of the Integrated Agreement for Lean Project Delivery 188
    focused, market-drive design 183
    higher quality, lower cost 181, *181*
    moving project workflow into program workflow 186
    whole team involvement and BIM 186, *187*
next generation
  breaking into new markets 139–43
    align facility and business strategies 141
    create competitive selling teams 140
    create new integrated project alliances 141
    creating competitive cultures 139
    deal effectively with emotional drivers 141
    exploit professional joint ventures 140
    generating own leads 139–40
    undertake rigorous pre-competition analysis 140
  development and exploitation of new categories 86–8
  opportunity mining 86, *87*
Nook, Greg (Executive VP, JE Dunn Construction) 134, 193

obsessive improvement, culture of 7
Ogilvy & Mather 146, 193
Ogilvy, David (Founder, Ogilvy & Mather) 146, 193
opportunity mining 86, *87*
organizations, divergent, factors in blending together 72, 74

Parsons (Houston) 135, 180, 185, 194

Peck, Chris (VP, McCarthy Building) 16, 193
Penland, Will (Retired Senior MD, CB Richard Ellis) 72, 74, 193
performance characteristics for outperforming competition 4–8
Perkins & Will 192
  founders' vision of the practice 12
persistent branding 6, *89*, 90–119
  revised thinking 91
  a summary 119
  terminology 92–6
  two levels of 91
  what is so important about a brand 99
  your brand 90
  see also branding
personal networks/networking 108, *109*, 126, 129
Porter & Hedges, LLP (Houston) 176, 192
Pringle Brandon Consulting (London, England) 34, 193
Pringle, Jack (President, RIBA 2005–2007) 193
  focused team of professionals from diverse backgrounds 34–5
professionals
  divergent disciplines, integrated teams 44–5
  from varied cultural backgrounds, focused teams 34–5
project delivery, sequential, no longer to be tolerated 166
project delivery, innovation and advancement imperative 178
Project Production Systems Lab., University of California (Berkley) 180, 188, 191
project success, divergent attitudes to 176–7
project teams 24, 27
public perception, manipulation of 114

Ratcliff Architects (San Francisco) 11, 193
Ratcliff, Christopher (Kit) 11, 193
Reed, Scott (at Cannon Design) 108, 194
Reliant Stadium 69
rewards 21, 23, 113
  evaluation and reward *38*, 39–40
  for perception as a high status company, substantial 46
  public 38, 112, 119, 128, 139
Rice University Building Institute (RBI) vii, xii–xiii, 166
  internet-based international survey of integrated project delivery 167–73
  operational centers xiii
  tomorrow's clients will demand teams of professionals 141
Rice University (Houston) 180, 184, 191
Robson, Thomas H. (Senior VP, HOK) 194
  clients expect us to be experts 80
Roche, James, and General Motors 12
Rogers, Peter (Director, Stanhope PLC) 83, 194

Rogers Stirk Harbour & Partners (London, England) 109, 193
Rollo, Mike (General Manager, Business Services, Leighton Contractors) 74, 170, 194
rotation, moving unique activities to continual activity 185
Rowlett, John (Joint Founder of CRS) 78
Royal Australian Institute of Architects 107
RTKL Associates (Chicago) 40–1, 192

Schmitt, Don (Principal, Diamond + Schmitt Architects) 33, 34, 194
Scott, Wally (Joint Founder of CRS) 78
Shepherd, Nick (Managing Partner, Drivers Jonas) 68, 83, 194
Simpson, Scott FAIA (Senior Principal, Klingstubbins) 129, 136, 167, 180, 194
  Higher quality, lower Cost 181, *181*
Skanska (Stockholm, Sweden) 80, 192
  a multi-local company 56
  values all offices must actively pursue 35
Skerman, Ben (Engineer, Worley Parsons) 194
  pursues brand extension projects 56
Smith Seckman Reid, Inc. 150, 151, 161, 180, 186, *187*, 193
Social Identification Theory 36, *37*
Society of Competitive Intelligence Professionals (Alexandria, Virginia) 105, 193
sociograms 135–6
solutions, seeking improvement 75
Stanhope PLC (London, England) 83, 194
Stewart, R.K. FAIA, Hon FRAIC (Principal, Gensler, President AIA (2007)) 143, 194
Stillman, Laura (Executive VP, Flad Architects) 174, 194
Stowkowy, Al (President and COO, Stuart Olsen) 100, 102, 150, 194
strategic alliances, new 7–8
Stuart Olsen 95, 100–2, 194
  constituents and messages designed for them 102
Stuart Olsen Construction 150, 154, 192
success, in science and industry vii
Sun Tzu, *The Art of War* 106
Sutter Health (Sacramento) 27, 31, 180, 188, 191, 193
Syska & Hennessy (New York) 95, 118, 157, 191

Tellepsen Builders (Houston) 132, 194
Tellepsen, Howard (Chairman, Tellepsen Builders) 132, 194
The Shaw Group 95
Thomsen, Charles B. FAIA, FCMAA (Former Chairman, 3D/I & Advisory Director, Parsons) 135, 180, 194
  Moving project workflow into program workflow 185
trading zones vii–viii

Trione & Gordon (Houston)
  formation of 70–1
  in talks with CB Richard Ellis Inc. 72, 74
Trione, Gerry (now Executive VP, CB Richard Ellis) 194
Trione, Gerry (Part Owner, Trione & Gordon) 70–1
Turner Construction 95, 193
  work with high potential employees 155
Tzannes, Alec (President, Royal Australian Institute of Architects) 107, 194
Tzannes Associates (Sydney, Australia) 194

United States Postal Service, value integrated branding event 45–6

values *29*, 30–46
  committing to 41–2
    at expense of a prestigious project 40–1
  contributing to competitiveness 5–6
  long-held, integration into a complex venture 34
  market leaders, dealing with values today 31–42
  steps to encourage a robust system of 38–40
    evaluation and reward *38*, 39–40
  working definition 30
vision 5, *9*, 10–27
  co-creation of 16
  engaging, measurable benefits 26–7
    better employees 26
    development process 27
    marketing strategy with a more cohesive face 26
    may increase competitiveness 27
    project teams 27
    sales, differentiating your company 26
  failures at creating a vision *14*
  impediments to becoming vision-driven 23
    communication 20
    expectations about 'buying in' 21
    lagger problem 21–3
    measurable 20
    occurrence of repetitive failure 19, *20*
    personal 20
  most effective practices 13–14
  selecting a CEO *9*, 10
vision-driven, working definition of meaning 23–6
  project teams in design and building industries 24–6
    company vision in forefront of consciousness 24
    sharing overarching sense of purpose 24, *25*
    vision relates to all key workers' aspirations 26
    vision translated into measurable goals 26
visions statements, effective
  characteristics 10–14
    addresses the future 10–11
    describes a meaningful, deeper purpose 12
    emotionally compelling 11

visions statements, effective (*continued*)
characteristics (*continued*)
    permeates everyday work 12–14, *13*
    reality based 12
  development of 15–19
    announce 15
    collaborate 16–19
    consult 16
    measuring commitment 15

Walter P. Moore (Houston) 69, 193
  designed baseball stadium with moveable roof 68
  structural engineering for the Astrodome 68
  world's largest retractable roofed football stadium 69
Willmott Dixon Construction (London, England) 50, 193
Worley Parsons (Queensland, Australia) 56, 194
Wyatt, Scott (Managing Principal, NBBJ) 18–19, 194

Yazdani Studio, Cannon Design 156